Signals and Communication Technology

More information about this series at http://www.springer.com/series/4748

Nan Chi

LED-Based Visible Light Communications

Nan Chi
Fudan University
Shanghai
China

ISSN 1860-4862   ISSN 1860-4870 (electronic)
Signals and Communication Technology
ISBN 978-3-662-58590-0   ISBN 978-3-662-56660-2 (eBook)
https://doi.org/10.1007/978-3-662-56660-2

Jointly published with Tsinghua University Press, Beijing, China

The print edition is not for sale in China Mainland. Customers from China Mainland please order the
print book from: Tsinghua University Press.
ISBN of the China Mainland edition: 978-7-302-33780-5

Preface

The emergence of "smart home" and the rapid spread of intelligent devices have made revolutionary changes to the category of the mobile digital terminal, which brings about a big test to the traditional access of network technology. The dilemmas of the "last mile" from optical fibers to home relate to the limited spectrum of resources regarding the wireless access network, the immaturity of ROF technologies, and electromagnetic radiation, which all restrict the bottleneck breakthrough. The world today is experiencing a profound revolution of access technology called "Anywhere, Anytime," and society is calling for a new access method, which can broaden the current spectrum resources, be "greener," as well as removable. As a result, visible light communication (VLC) has emerged, as modern times require.

Visible light communication uses white LEDs as a light source and utilizes high-speed flashing signals carried by the LED lights to transmit information. Visible light communication is the result of the combination of both lighting and communication characteristics. Because of its great advantages, such as high efficiency, low cost, and a long lifetime, it is certain that LEDs will become our main sources of lighting, instead of incandescent bulbs and fluorescent lamps. In 2011, our country launched incandescent phase-out programs and, thus, planned to completely ban the sale of common lighting incandescent lamps by 2016. There is no doubt that LEDs will become the next generation of lighting technology, which has been a recent trend in *The Times Magazine*. The popularity of solid-state lighting makes the light source of VLC available everywhere. Standing on the shoulders of giants, visible light communication is developing rapidly with the boom of the LED industry. Due to the LED's features, such as energy saving and cost saving, visible light communication will serve as a new means of "green" communication and will make a great contribution to the country's energy conservation plan.

Visible light communication has the following positive characteristics. To start, white light is safe to human eyes, and the power of indoor white LED lamps can reach up to ten watts or more, which means visible light communication has a very high signal-to-noise ratio, with greater bandwidth potential. Second, visible light communication technology has no electromagnetic pollution, so it can be applied to aircrafts, hospitals, industrial controls, and other RF-sensitive areas. In addition, visible light communication combines illumination, communication, positioning, as well as other functions, with low energy consumption, less equipment, and other advantages, which meets national energy conservation strategies. The fourth advantage is that, since visible light communication uses an unlicensed spectrum, its applications are flexible, and it can be used alone or as a valid backup for RF wireless devices. Furthermore, visible light communication is suitable for information security applications. As long as there are obstacles that visible light cannot penetrate, information within the illumination network will not be leaked, so visible light communication has a high degree of confidentiality.

Since the concept of visible light communication was proposed in 2000, it has quickly gained attention and support from all over the world. In just ten years, it has developed rapidly. The transmission rates have improved from tens of Mbit/s to 500 Mbit/s and even to 800 Mbit/s. In addition, VLC technology has developed rapidly from off-line to real time, from low-end modulation to high-order modulation, from point-to-point to multiple-input multiple-output (MIMO). VLC technology has impacted the global market, and *Times Magazine* rated it as one of the "Top 50 Worldwide Scientific and Technological Inventions" in 2011. Thus, today's VLC technology research is experiencing increasingly active development, where new concepts and new technologies are emerging endlessly. Whether from the national strategic level, or as an urgent need for energy conservation, or just by considering the huge market potential, VLC is making a huge impact within China. As a combination of both new methods of illumination and optical communication, VLC is promoting the development of the next generation of lighting, as well as an access network, and represents great technological progress, which has lead to it becoming one of the focuses and key points of international competition.

The author Nan Chi is a Professor and Doctoral Mentor of the Communications Department of the School of Information at Fudan University. She is a Senior Member of the Optical Society of America (OSA) and Member of Institute of Electrical and Electronics Engineers (IEEE), Technical Committee on Integrated Optoelectronics of Chinese Optical Society, as well as the Optical Communications Committee of China Institute of Communications. She has won the New Century Excellent Talents of the Ministry of Education, Shanghai Shuguang scholars, Japan Okawa intelligence funds, Shanghai Pujiang Talents, and Shanghai's top ten cutting-edge IT. Additionally, she has undertaken a number of national projects, including 973 project topics, 863 projects and Natural Science Foundation projects. Furthermore, she has published more than 300 papers, including more than 200 SCI papers, which have been cited more than 2000 times. Her research interests are in the areas of coherent optical transmission, visible light communication, and optical packet/burst switching.

Currently, there are no domestic books that systematically introduce white LED-based visible light communication. This book fills the blanks and provides a more detailed description on the visible light communication system, which can be used as a teaching book for university students or a fundamental reference for engineers and other technical staff. In this book, Chap. 1 introduces the basic concepts of visible light, which also provides a background and some primary knowledge. Chapters 2–6 discuss the foundation of visible light technology, which introduces the transmitting parts, channel models, receiving parts, as well as modulation and equalization techniques. In order to help readers have a deeper understanding of visible light technology, Chap. 7 provides a few visible light communication system experiments. Chapter 8 focuses on the upper layer protocols of the visible light communication system, and finally, Chap. 9 has information about our future predictions regarding the development trends of the visible light communication system.

The composition of this book has obtained great help from teachers and students of the Shanghai Science and Technology project. Among them, teachers Muqing Liu and Xiali Zhou wrote part of Chap. 2; teachers Xinyue Guo and Minglun Zhang wrote the channel model portion of Chap. 3; teachers Yonggang Zhang and Shaowei Wang wrote the detector portion of Chap. 4; and teacher, Rui Zhang, wrote part of Chap. 8. The author also thanks the support and help from students like Rongling Li, Yuanquan Wang, Yiguang Wang, Xingxing Huang, Jiehui Li, and Chao Yang and also Allison Lasley for assisting with English translations. The composition of this book was written relatively in a hurry, so inadequates are inevitable. We sincerely hope to receive valuable suggestions from readers for future improvements and enhancements.

Shanghai, China   Nan Chi

Contents

Chapter 1
Outline

1.1 Introduction

The emergence of "smart home" and the rapid spread of intelligent devices have made revolutionary changes to the category of the mobile digital terminal, which brings about a big test to the traditional access of network technology. The dilemmas of the "last mile" from optical fibers to home relate to the limited spectrum of resources regarding the wireless access network, the immaturity of ROF technologies, and electromagnetic radiation, which all restrict the bottleneck breakthrough. The world today is experiencing a profound revolution of access technology called "Anywhere, Anytime," and society is calling for a new access method, which can broaden the current spectrum resources, be "greener," as well as removable. As a result, visible light communication (VLC) has emerged, as modern times require.

The concept of VLC was born in the year 2000. Employing light-emitting diodes (LED) as a light source, the VLC system can provide high-speed communication as well as lighting. Nowadays, white LED has been widely used in different fields including signal transmission, display, light, etc. Compared with other light sources, white LED has a higher modulation bandwidth. What's more, it has good modulation performance and high response sensitivity. Thus, through LEDs, the signal can be modulated into visible light. White LEDs, combining both light and data transmissions, have further promoted the development of VLC technologies.

Nowadays, wireless spectrum resources are tight since many bands are already occupied, as shown in Fig. 1.1. Visible light employed by the VLC system is still blank within the spectrum and can be used without authorization. Thus, VLC technology can use the blank spectrum and the available resources successfully and efficiently. This has expanded the spectrum of the next generation's use of wideband communication. In addition, compared with other wireless technologies, VLC

© Tsinghua University Press, Beijing and Springer-Verlag GmbH Germany 2018
N. Chi, *LED-Based Visible Light Communications*, Signals and Communication
Technology, https://doi.org/10.1007/978-3-662-56660-2_1

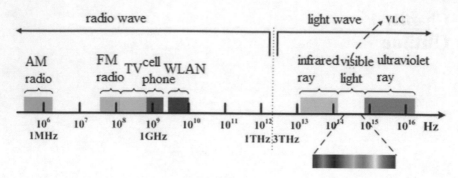

Fig. 1.1 Schematic diagram of spectrum resources

also has many other advantages, such as high security, good confidentiality [1]. The purpose of this chapter is to introduce the basic concepts of VLC and provide some background information, as well as preliminary knowledge.

1.2 LED Market Trends

The twenty-first century is the era of the LEDs, which are used as light sources in VLC. The advent of LEDs was in the 1960s, and since then, the science has developed rapidly. In a just few decades, the light has transformed from a single color to multiple colors, from low luminance to high luminance, increased in lifespan longevity, and the market size, which started out small, has increased dramatically. Since the birth of LEDs, every 10 years the luminance has increased by 20 times, while the price has decreased to 1% of the original price. The continuous improvement of this technology and the enrichment of its functions have brought dramatic changes to human society. Its influence has penetrated into the realm of global science and technology, economics, life philosophy, and other fields. However, it has particularly influenced the field of lighting, with a powerful advantage and competitiveness. Compared with incandescent bulbs and energy-saving lamps, LEDs have numerous benefits such as high efficiency, low price, long service life. A LED's power consumption is only 1/10 of incandescent bulbs' and 1/4 of energy-saving lamps. Additionally, its luminous efficiency is higher and can reach up to 249 lm/w, which is about 4 times of fluorescent lamp, and its life span can reach up to 100,000 h, which only uses 1/1000 of rare earth elements, compared to conventional energy-saving lamps which use much more. These unbeatable properties have lead to the LEDs' quick occupation within the global market and its increasing favoritism by countries all around the world. Countries have successively launched incandescent phase-out programs, as seen in Table 1.1, and traditional lighting technology is rapidly evolving into solid-state lighting technology. The LED market share is shown in Fig. 1.2. There is no doubt

Table 1.1 Different countries and their replacement of incandescent lamp procedures

Continent	Country	Incandescent phase-out programs
Asian	China	2011.10.1–2012.9.30: Transitional period 2012.10.1: Ban on import and sales of 100 w and above general lighting incandescent lamps 2014.10.1: Ban on import and sales of 60 w and above general lighting incandescent lamps 2015.10.1–2016.9.30: Mid-term evaluation period 2016.10.1: Ban on import and sales of 15 w and above general lighting incandescent lamps
	India	Before 2010: Replace 400 million incandescent lamps with energy-saving lamps
	The Philippines	After 2010: Ban on use of incandescent lamps
	Malaysia	After 2014: Stop the production, import and sales of incandescent lamps
Europe	EU	After 2016: Stop using incandescent and halogen lamps
	Ireland	After 2012: Stop using incandescent lamps
	Switzerland	Ban the use of F- and G-level incandescent lamps
	UK	After 2011: Stop using incandescent lamps
America	Canada	After 2012: Ban on use of incandescent lamps
	USA	After 2020: Ban on use of incandescent lamps under 45 lm/w
	Cuba	After 2005: Ban on import of incandescent lamps, taking energy-saving lamps for replacement
Oceania	Australia	After 2010: Ban on sales of incandescent lamps

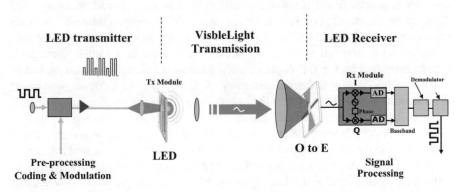

Fig. 1.2 Block diagram of the visible light communication system

that LEDs will become the next generation of lighting technology, which has been a recent trend in *The Times Magazine*. The popularity of solid-state lighting makes the light source of VLC available everywhere. Standing on the shoulders of giants, VLC is developing rapidly with the boom of the LED industry.

Overall, LED's market share has grown steadily in recent years. According to the analysis report of LED market released by Chinese industry research network, LED application in our country is the fastest growing part of the LED industry chain; the overall growth rate was close to 38% in 2014. Among them, the general lighting market growth rate is about 69%, accounting for 41% in the domestic market. LED backlight application growth rate is slow, and the annual growth rate is about 17%.

In 2014, domestic LED display applications also grew rapidly, with annual growth rate of about 35%. In addition, the applications of LED automotive lighting, medicine, agriculture, and other emerging fields are also expanding. The applications of smart lighting, light communication, and wearable devices are the new highlights of the LED application in the future.

1.3 The History of Visible Light Communication

VLC technology based on white LED provides an additional service at a comparably low cost and combines both lighting and communication characteristics. Besides, it is available worldwide, free from electromagnetic interference, and "green." Because of these advantages of VLC technology, it has received prompt attention and support around the world since its advent. Since its inception to now, which has been just a few dozen years, it has experienced continuous breakthroughs.

Japanese researchers first proposed the concept of VLC. In 2000, Japanese researchers proposed and simulated an indoor communication system, using LED lights as the communication base station, to transmit information through a wireless transmission. From then on, the Japanese researchers saw the development prospects of VLC and put a lot of effort into their research. In 2009, Nakagawa Laboratories proposed a full-duplex multiple access VLC system based on carrier sense multiple access/collision detection (CSMA/CD), to achieve high-speed communication reaching 100 Mbit/s. In order to achieve the practicability of VLC, "Visible Light Communications Consortium" (VLCC) was established in Japan in 2003 and quickly became an international organization. Japanese research about VLC technology has developed from a transmission system into multiple applications. Researchers have proposed that VLC technology should be applied to billboards, lighthouses, positioning systems, intelligent transportation systems, and so on. In 2008, researchers conducted an experiment in Kujukuri-machi, Japan, which used the LED of a lighthouse as a transmitter and an image sensor as a receiver. With this experiment, they realized a transmission rate of 1022 bit/s at the farthest distance of 2 km. In 2009, VLCC exhibited a digital billboard sample applying VLC technology. The billboard used its backlit LED to transmit data so the users could download the information needed. In 2010, VLCC and the Japan

Traffic Management Technology Association conducted a VLC experiment that used an image sensor as the receiver and LED traffic lights as the transmitter. With this experiment, they successfully achieved the transmission rate of 4800 kbit/s at a distance of 300 m. In 2012, Keio University researchers created a voice-assisted system for visually impaired persons, using a smartphone to detect the user's location information and then guide them. Thus said, Japan had taken a big step in the development of products based on VLC technology. In 2012, Casio developed a new Apple application "Picapicamera" where users could instantly share photographs using VLC technology.

Although the birthplace of visible light technology is in Japan, researchers in USA and EU are not far behind. Due to the government's attention and ample research funds, they have attained many outstanding achievements. In 2008, the EU launched the OMEGA project to develop research on 1 Gbit/s, or the ultra-high-speed home access network. VLC wireless communication technology is the focus of the research. The theoretical speed of the test network was 1.25 Gbit/s, and the maximum transmitting speed was 300 Mbit/s. In 2008, the US National Science Foundation carried out "Smart Lighting Communications" (SLC) projects, mainly for VLC wireless communication technology research. In 2011, Germany, Norway, Israel, and the USA co-founded Li-Fi Alliance and conducted Internet research for aerospace systems, which used VLC technology for a wireless network environment when flying. In 2012, funded by the UK Engineering and Physical Sciences Research Council (EPSRC), the UK and USA scientists carried out an "ultra-parallel visible light communication" (UP-VLC) project to explore the implementation of free space and spatial multiplexing of guided wave VLC. The Berlin Heinrich Hertz Institute, in cooperation with Siemens, conducted high-speed VLC technology research in 2012. They used discrete multitone (DMT) modulation technology, a RGB-LED transmitter, and a PIN-based receiver to achieve a single-channel (red optical channel) transmission rate of 806 Mbit/s.

We have made some comprehensive statistics about breakthroughs in VLC transmission experiments in recent years, as shown in Table 1.2. VLC technology has developed rapidly in just ten years. The transmission rates have been promoted from tens of Mbit/s to 500 Mbit/s, and even further to 800 Mbit/s. Now, the rates have exceeded Gbit/s, and higher speed communications have also been in sight. In addition, VLC technology has developed rapidly from off-line to real-time, from low-end modulation to high-order modulation, from point-to-point to multiple-input multiple-output (MIMO). VLC technology has impacted the global market, and *Times Magazine* rated it as one of the "Top 50 Worldwide Scientific and Technological Inventions" in 2011. Thus, today's VLC technology research is developing rapidly, where new concepts and new technologies are emerging endlessly.

Table 1.2 Summary of the transmission rates in the VLC system test

Channel	Equalization	Modulation scheme	Demonstrated data rate	Receiver	Distance	Research institute
White channel		OOK	1022 bit/s	Image sensor	2 km	VLCC
White channel	Pre	OOK	40 Mbit/s BER < 10^{-6}	PIN	2 m	University of Oxford
Blue channel	Pre	OOK	80 Mbit/s BER < 10^{-6}	PIN	10 cm	University of Oxford
White channel	Post	DMT	101 Mbit/s	PIN	1 cm	Fraunhofer HHI
White channel	Post	DMT	230 Mbit/s	PIN	70 cm	Fraunhofer HHI
White channel	Post	DMT	513 Mbit/s	APD	30 m	Fraunhofer HHI
RGB-LED	Post	DMT	803 Mb/s	APD	12 cm	Fraunhofer HHI
RGB-LED	Post	DMT	1.25 Gb/s	APD	10 cm	Fraunhofer HHI
RGB-LED	Post	DMT	2.1 Gb/s	APD	10 cm	Scuola Internazionale Superiore di Studi Advanzat
RGB-LED	Post	DMT	2.1 Gb/s	APD	10 cm	Scuola Internazionale Superiore di Studi Advanzat
White channel	Post	CAP	1.1 Gb/s	PIN	23 cm	National Chiao Tung University
RGB-LED	Post	CAP	3.22 Gb/s	PIN	25 cm	National Chiao Tung University
RGBY LED	Post	DMT	5.6 Gb/s	PIN	1.5 cm	Scuola Internazionale Superiore di Studi Advanzat
RGB-LED	Pre/post	SC	4.22 Gb/s	APD	1 cm	Fudan University
RGB LED	Pre/post	CAP	4.5 Gb/s	PIN	2 m	Fudan University
RGB LED	Pre/post	CAP	8 Gb/s	PIN	1 m	Fudan University
uLED	Pre/post	PAM4	2 Gb/s	APD	60 cm	University of Cambridge
RGB LED	Pre/post	PAM8	3.375 Gb/s	PIN	1 m	Fudan University
RGBY LED	Pre/post	DMT	9.51 Gb/s	PIN	1 m	Fudan University
RGB LED	Pre/post	DMT	10.4 Gb/s	PIN	1.5 m	University of Oxford

1.4   The Composition of the Visible Light Communication System

The block diagram of a VLC system based on white LEDs is shown in Fig. 1.3. The system is composed of three parts: the LED transmitter, the visible light transmission, and the LED receiver. After pre-treatment and coded modulation, the original binary bit stream drives the LEDs and converts the electrical signals into optical signals with intensity modulation. Pre-treatment, namely pre-equalization, is used to compensate for any signal distortion due to other devices and channels. The adoption of pre-equalization technology can improve the response bandwidth of LEDs, as well as increase the transmission rate. In addition, post-equalization at the receiving end can compensate for other channel losses, such as phase noise. The principles of pre-equalization and post-equalization will be introduced in detail in Chap. 6. The purpose of coded modulation is to achieve a higher transmission rate on a limited bandwidth. In order to improve the transmission rate of the white LED communication system, which is limited by the VLC bandwidth, we can design and adopt high-order modulations and coding techniques to improve the spectral transmission efficiency. Currently, quadrature amplitude modulation–orthogonal frequency division multiplexing (QAM-OFDM) is the most widely adopted method by researchers.

There are currently two types of LEDs on the market, phosphor LEDs and RGB-LEDs, which are used as the light source of white LEDs in the VLC system. The phosphor LED is the most widely used type of LED, and its principle uses blue light to inspire the yellow phosphor to produce white light. This type of LED has a simple structure, lower cost, and a relatively lower modulation complexity, but the modulation bandwidth is very low, which makes the spectrum utilization rate also low. Since the modulation bandwidth of a phosphor LED is only a few megabytes, the transmission rate of the system is limited. When using on–off keying–not-return-to-zero (OOK-NRZ) modulation, the maximum transmission rate can reach 10 Mbit/s. This is caused by the low response speed of the yellow phosphor. Another common type of LED is the RGB-LED. Its purpose is to encapsulate the red, green, and blue LED chips and mix the light emitted to produce white light. The RGB-LED has a very high modulation bandwidth, which is a good sign that it will be used in future high-speed signal transmissions. However, the modulation complexity is relatively high, and some glitches, like how to control the three chips to avoid flickering and maintaining the mixed color stability, need to be further studied. VLC systems based on two types of LEDs have their respective advantages. The system based on phosphor LEDs can be achieved easily and has a lower cost, while the other based on RGB-LEDs can achieve a higher transmission rate.

The optical signal carrying the data is transmitted in free space and focused on a photodetector through the lens in front of the receiver. In this system, most of the received energy comes from the direct path of the line of sight. The photodetector converts the received optical signal into electrical signals, and the original transmitted signal is recovered after signal processing, demodulation, decoding, etc.

There are three types of receivers that can be used in VLC systems: PIN-based receivers, APD-based receivers, or an image sensor. The PIN-based receiver has a fast response speed, high sensitivity, and low cost. The APD-based receiver has a faster response speed, higher sensitivity, and a higher SNR, but the cost is relatively high. Therefore, the current high-speed VLC systems usually employ receivers based on PIN or APD. While for the receiver based on an image sensor, the response speed is slow and the sensitivity is also relatively low. However, it can receive data from multiple sources at the same time and the transmission distance is longer. Therefore, it is often used in MIMO-VLC systems, as well as many other applications, such as intelligent transportation systems, positioning systems.

1.5 Advantages of Visible Light Communication Technology

VLC, as an emerging wireless communication technology, is receiving more and more attention. Some of the main advantages are reflected in the following aspects.

First, due to the increasing scarcity of radio spectrum resources, the introduction of VLCs is a huge expansion of the communication spectrum. Because the use of more mobile digital terminals is in high demand, especially for the needs of users for "Anywhere, Anytime" video services, the current wireless spectrum resources will soon be exhausted. Therefore, we need to adopt new technologies to expand the wireless spectrum. Visible light has a huge bandwidth from 380 to 780 nm (equivalent 405 THz), and it can ease the immediate need that our radio spectrum resource is running short of.

Second, VLC technology uses LED lights that have high-speed modulation characteristics, not only to achieve lighting but also to realize Internet communication. At the same time, it can achieve intelligent control for controlling terminals, such as home appliances and security equipment. VLC technology uses fluorescent or light-emitting diodes that emit a fast flashing light to transmit information that cannot be seen by the human eye. Therefore, it is a type of "green" wireless communication technology and can operate without radiation damage to the human body. Besides, LED is widely recognized as an energy source. The organic integration of lighting, intelligent communication, and intelligent control can help humanity conserve more precious energy and promote a "green" lifestyle.

Another advantage is that a variety of VLC application scenarios can be used as an effective complement to the wireless communications that already exist. Due to the fact that VLC technology has no electromagnetic pollution, it can be applied to aircrafts, hospitals, industrial control, and other RF-sensitive areas. Due to the combination of both lighting and communication, it is suitable for the applications of a smart home, intelligent transportation, and other areas. It is also suitable for underwater high-bandwidth communication with its blue-green LED semiconductor lighting technology. In addition, it is also suitable for information security applications, as long as there is an obstacle blocking visible light and the information within the semiconductor lighting information network cannot be divulged.

Finally, VLC technology has the potential for high-speed communication, which may provide some technical support for future high-speed access. For example, because white light has many advantages, such as it is safe to human eyes, the sum of the interior white LED lights power is up to ten watts or more, and it creates VLC with a very high signal-to-noise ratio, these attributes create a perfect foundation for high-speed communication. Realistically, other technologies cannot be compared to it. Presently, the speed rate of VLC experiments has reached 3.4 Gbit/s.

Because VLC technology has several advantages as shown above, researchers around the world are devoting a great deal of enthusiasm to it and are making rapid advances in VLC technology. There are new breakthroughs almost every day. VLC technology has diverse applications, a huge number of users, and good prospects for development, so the practical use of VLC technology is worth the wait.

1.6 Research Trends

Currently, major foreign equipment manufacturers, universities, research institutes, and so on are conducting research on VLC technology. VLC technology has become an international research hot spot. However, the development of VLC technology also has some limitations. The most important challenge is the limited bandwidth of white LEDs, which limits the transmission rate. The modulation bandwidth of phosphor LEDs, currently most widely used, is only a few megabytes. Therefore, how to improve the LED modulation bandwidth and its system transmission rate has become a key point of research.

To begin, the researchers add a blue light filter before the signal detection to filter out the slow response of the yellow light component, which will make the phosphor LED's modulation bandwidth increasing from 3 to 10 MHz. Then, using equalization technology to adjust the LED's frequency response, we can increase the bandwidth to tens of megabytes. If using the RGB-LED instead of the phosphor LED as the light source, a higher modulation bandwidth can also be obtained. At the same time, through the use of WDM techniques, the system transmission rate can be improved. Using the MIMO technology, through spatial multiplexing, can also improve the system transmission rate. By using a higher-order modulation scheme and DMT technology, the system transfer rate can be improved even further.

Blu-ray filtering and equalization techniques are easy to realize and can increase the phosphor LED's modulation bandwidth. To a certain extent, they can also improve the system's transmission rate. WDM technology is only applicable to a VLC system that uses a RGB-LED as the light source. Since a VLC system using a RGB-LED has some unique characteristics, such as a high modulation bandwidth and three kinds of emitted monochromatic light, the transmission rate can be significantly improved. To further improve the rate, we need a higher-order modulation format, such as QAM-DMT, but this will further increase the complexity of the system. If we use MIMO technology, since the image detector limits its transmission rate, the current rate achieved is not very high. However, it is the most promising technique. Unfortunately, by using some high-order modulation formats to improve the system's transmission rate, such as increasing the modulation order, using a more complex system, and increasing the receiver sensitivity requirements, it is quite inevitable to encounter a problem. However, through spatial multiplexing, MIMO technology can achieve high-speed communications in the limited bandwidth. Therefore, with the development of technology, MIMO technology will be a powerful option for the future of the high-speed VLC system.

1.7 Brief Summary

This chapter mainly introduces some basic concepts and background knowledge about VLC in order to help readers lay a foundation to further understand the up and coming VLC technologies. First, the chapter briefly describes the LED market trends and shows readers its strong market potential. Next, the chapter introduces the development of VLC, which helps the readers understand the development process of VLC technology, from the birth of its concepts to the recent Gbit/s transmission rate breakthroughs. Then, the third section introduces the basic components of a VLC system as well as provides a brief description of each part. The last part is the sum and future prospects of VLC technology research trends.

References

1. Chi, N., Haas, H., Kavehrad, M., Little, M.T., Huang, X.: Visible light communications: demand factors, benefits and opportunities. IEEE J. Wirel. Commun. **22**(2), 5–7 (2015)
2. Langer, K.-D., Vučić, J., Kottke, C., et al.: Advances and prospects in high-speed information broadcast using phosphorescent white-light LEDs. In: ICTON, Mo.B5.3 (2009)
3. Cui, K., Chen, G., Xu, Z., Roberts, R.D.: Line-of-sight visible light communication system design and demonstration. CSNDSP 2010, OWC-21 (2010)
4. Tanaka, Y., Haruyama, S., Nakagawa, M.: Wireless optical transmissions with white colored LED for wireless home links. Indoor Mobile Radio Commun. **2**, 1325–1329 (2000)
5. Le Minh, H., O'Brien, D., Faulkner, G., et al.: High-speed visible light communications using multiple-resonant equalization. IEEE Photonics Technol. Lett. **20**(14), 1243–1245 (2008)
6. Le Minh, H., O'Brien, D., Faulkner, G., et al.: 80 Mbit/s visible light communications using pre-equalized white LED. ECOC 2008, P.6.09 (2008)
7. Vučic´, J., Kottke, C., Nerreter, S., et al.: 513 Mbit/s visible light communications link based on DMT-modulation of a white LED. J. Lightwave Technol. **28**(24), 3512–3518 (2010)
8. Kottke, C., Habel, K., Grobe, L., et al.: Single-channel wireless transmission at 806 Mbit/s using a white-light LED and a PIN-based receiver. ICTON, We.B4.1 (2012)

9. Khalid, A.M., Cossu, G., Corsini, R., et al.: 1-Gb/s transmission over a phosphorescent white LED by using rate-adaptive discrete multitone modulation. IEEE Photonics J. **4**(5), 1465–1473 (2012)

10. Wu, F.-M., Lin, C.-T., Wei, C.-C., Chen, C.-W., Huang, H.-T., Ho, C.-H.: 1.1-Gb/s white-LED-based visible light communication employing carrier-less amplitude and phase modulation. IEEE Photonics Technol. Lett. **24**(19), 1730–1732 (2012)

11. Grubor, J., Lee, S.C.J., Langer, K.-D., Koonen, T., Walewski, J.W.: Wireless high-speed data transmission with phosphorescent white light LEDs. ECOC 2007, 1–2 (2007)

12. Park, S.-B., Jung, D.K., Shin, H.S., Shin, D.J., Hyun, Y.-J., Lee, K., Oh, Y.J.: Information broadcasting system based on visible light signboard. Proc. Wireless Opt. Commun. **2007**, 311–313 (2007)

13. Le Minh, H., O'Brien, D., Faulkner, G., et al.: 100-Mb/s NRZ visible light communications using a postequalized white LED. IEEE Photonics Technol. Lett. **21**(15), 1063–1065 (2009)

14. Azhar, A.H., Tran, T.A., O'Brien, D.: Demonstration of high-speed data transmission using MIMO-OFDM visible light communications. In: IEEE Globecom 2010 Workshop on Optical Wireless Communications, pp. 1052–1056 (2010)

15. Vučić, J., Kottke, C., Nerreter, S., Büttner, A., Langer, K.-D., Walewski, J.W.: White light wireless transmission at 200+Mb/s NetData rate by use of discrete-multitone modulation. IEEE Photonics Technol. Lett. **21**(20), 1511–1513 (2009)

16. Vučić, J., Kottke, C., Habel, K., Langer, K.-D.: 803 Mbit/s visible light WDM link based on DMT modulation of a single RGB LED luminary. OSA/OFC/NFOEC: OWB6 (2011)

17. Khan, T.A., Tahir, M., Usman, A.: Visible light communication using wavelength division multiplexing for smart spaces. In: 2012 IEEE Consumer Communications and Networking Conference, pp. 230–234 (2012)

18. Zeng, L., O'Brien, D.C., Le Minh, H., et al.: High data rate multiple input multiple output (MIMO) optical wireless communications using white LED lighting. IEEE J. Sel. Areas Commun. **27**(9), 1654–1662 (2009)

19. O'Brien, D.: Optical multi-input multi-output systems for short-range free-space data transmission. CSNDSP, pp. 517–521 (2010)

20. O'Brien, D.: Multi-input multi-output (MIMO) indoor optical wireless communications. In: 2009 Conference on Signals, Systems and Computers, pp. 1636–1639 (2009)

21. Wang, Y., Zhang, M., Wang, Y., et al.: Experimental demonstration of visible light communication based on sub-carrier multiplexing of multiple input single output OFDM. In: OECC, pp. 745–746 (2012)

22. Lin, X., Ikawa, K., Hirohashi, K.: High-speed full-duplex multiaccess system for LED-based wireless communications using visible light. In: International symposium on optical engineering and photonic technology: OEPT 2009, pp. 1–5 (2009)

23. Zeng, L., et al.: Improvement of date rate by using equalization in an indoor visible light communication system. In: ICCSC, pp. 678–682 (2008)

24. Shrestha, N., Sohail, M., Viphavakit, C., et al.: Demonstration of visible light communications using RGB LEDs in an indoor environment. In: ECTI-CON, pp. 1159–1163 (2010)

25. Cossu, G., Khalid, A.M., Choudhury, P., Corsini, R., Ciaramella, E.: 3.4 Gbit/s visible optical wireless transmission based on RGB LED. Opt. Express **20**, B501–B506 (2012)

Chapter 2
The Transmitter of the Visible Light Communication System

In recent years, semiconductor lighting technology, also known as green lighting, has been developed rapidly. The light-emitting diode (LED) is one of the most promising solid green light sources of the twenty-first century. LED communication offers an entirely new paradigm within wireless technology in terms of communication speed, flexibility, usability, and security. Right now, a great opportunity is available to us by using the current state-of-the-art LED lighting solutions for illumination and communication simultaneously, as well as with the same modulation. This can be done due to the ability that we can modulate LEDs at speeds far faster than the human eye can detect, while still providing artificial lighting. Thus, while LEDs will be primarily used for illumination, their secondary duty could be data communication through a lighting system. Visible light communication (VLC) technology, with visible light as the carrier, is a new type of wireless optical communication technology based on the LED light system. In this chapter, we will introduce the different kinds of LED, their driven design, their lighting light field, and their visual design.

2.1 Summary of the LED

2.1.1 The Development of the LED Light Source

Light-emitting diodes were discovered early in the twentieth century. Even as early as 1907, the luminescence phenomenon, produced by the PN junction of silicon carbide (SiC), was discovered. In the late 1920s and early 1930s, another discovery was made, which was that yellow phosphor derived from zinc sulfide and copper also emitted light. However, one after the other, people have stopped researching this science because the light produced by SiC and yellow phosphor is very dark. Nevertheless, the early research left an everlasting impression of electroluminescence on people.

© Tsinghua University Press, Beijing and Springer-Verlag GmbH Germany 2018   13
N. Chi, *LED-Based Visible Light Communications*, Signals and Communication
Technology, https://doi.org/10.1007/978-3-662-56660-2_2

In 1962, Dr. Nick Holonyak invented the red LED with semiconductor compound materials of red phosphor gallium arsenide (GaAsP), which was developed by a joint laboratory effort including General Electric (GE), Monsanto, and International Business Machines (IBM). This LED was used as an indicator, until the luminescence efficiency of red LED decreased to 0.1 lm/W in 1965. In 1968, people made the luminescence efficiency of the GaAsP LED improve significantly, up to 1 lm/W, through a doping process. Furthermore, they also fabricated the first orange and yellow LEDs. The industry generally suggests that the red LED, which has real commercial value, appeared in the late 1960s.

From 1970 to 1971, the green LED, whose luminescence efficiency corresponded to that of the red LED, gradually appeared. Now entering into the mid-1970s, the gallium phosphide (GaP) green LED was born. Before entering the 1980s, LED colors were primarily limited to just three colors, red, yellow, and green. The luminous intensity of LEDs remained at the level of 1 mcd, and the luminescence efficiency was around 1 lm/W. LED lamps were mainly used as indicator lamps for the electronic products in this period. After entering the 1980s, researchers tried to use aluminum gallium arsenic (AlGaAs) to improve the LED's performance, and thus, the green and orange LEDs were fabricated. This resulted in an increase in the luminescence efficiency of red LEDs by nearly ten times. Because of the improvement of the LEDs' luminescence efficiency, LEDs were then applied to many more areas, such as outdoor display systems, bar code systems, photo-electric conduction systems and medical equipment.

On another note, the idea of a white light-emitting diode (LED) is not inherently unusual or surprising. It was not until the recent successful creation of high-frequency blue/violet light LEDs that the white LED made its debut. With the attributes of a highly efficient and cold light source, the white LED market will surely have enormous growth. In addition, the white LED will be a promising candidate for the replacement of the light bulb once production costs fall thanks to the future's technology advances. Nichia Co., who also created the first blue LED, produced the first commercially available white LED that was based on phosphors. Nichia used a blue light-emitting gallium indium nitride and coated the chip with the yellow fluorescent phosphor Y3Al5O12: Ce, well known now as (YAG: Ce). In addition, yellow phosphor Tb3Al5O12: Ce (TAG: Ce) was also used by OSRAM Co. with the GaInN chip for white light production. YAG and TAG, when activated with trivalent cerium, are efficient phosphors for converting the blue LED radiation into a very broad yellow emission band. The yellow emission from YAG: Ce and TAG: Ce is intense enough to complement the residual blue light which escapes through the phosphor in order to produce a white light. By far, YAG: Ce and TAG: Ce are the most excellent phosphors that have been applied successfully in the white phosphor-based LED commercial market. Therefore, any improvement in the luminescence of YAG: Ce and TAG: Ce is extremely valuable to improve the light efficiency of different applications.

The discovery of the white LED has great significance for LEDs as a new generation of an electric light source in the twenty-first century. White LEDs are commonly used as the back light source of mobile phones, PDA, and small LCD displays. The development of white LEDs is changing with each passing day. Similar to the microprocessor, LED development abides by Moore's law. Its brightness will double every 18 months, and its photosynthetic efficiency will be enhanced at an annual rate of 10–20 lm/W. The development of LEDs also follows Haitz law, which is named after Agilent former scientist Roland Haitz. This law states that the price of a LED will be one-tenth of the original price every 10 years, and its performance will enhance 20-fold. In recent years, the luminous efficiency of LEDs have enhanced unceasingly. Many of the related products' performances are significantly better than the affordable fluorescent lamp. For example, United States' Cree, Inc. launched a white LED with a luminous efficiency of 131 lm/W in 2006, and now, the luminous efficiency of white LEDs is already as high as 150 lm/W, thanks to Nichia Corporation. In January 2007, Philips Lumileds Lighting Company, who originally obtained a rate of 115 lm/W, promulgated the luminous efficiency of a high-power white LED. With the development of a fixed power LED, some general lighting products, such as the large-size LCD backlight, miner's lamps, LED road lighting lamps, and even lanterns, have been constantly emerging. Experts predict that the white light semiconductor lighting will create an output value of 150 billion RMB, and accumulatively save about 400 billion degrees of energy in China from 2005 to 2015.

2.1.2 The LED's Luminescence Mechanism

Luminescent materials, also called phosphors, are mostly solid inorganic materials consisting of a lattice host and usually intentionally doped with impurities. According to the mechanism of luminescence, an electrical light source used for lighting can be divided into several major categories including thermal radiation light sources, thermal gas discharge light sources, and electroluminescent light sources. Currently, the thermal radiation light source is represented by an incandescent bulb and the thermal gas discharge source is represented by a fluorescent lamp, which are both widely used. Meanwhile, electroluminescence is a type of light source that is currently being further developed. According to the principle of light, electroluminescent lighting can be divided into two categories: intrinsic electroluminescence and injected electroluminescence. For example, the semiconductor light-emitting diode is a type of an injected electroluminescent, which can be considered as a phenomenon that emits the solid material's light directly under the effect of the electric field.

As the name implies, a semiconductor diode is subject to a small forward-biased voltage, where the electrons are injected into a conduction band that is normally empty. By emitting their energy as photons, the injected electrons are recombined with the holes in the valence band. This process is called electroluminescence or spontaneous emission. Since an optical cavity does not need to provide photon feedback, and the emitted photons have random phases, the LEDs are considered incoherent light sources. Furthermore, the emitted photon energy is close to the band-gap of the semiconductor material. When a PN junction is forward biased, both electrons from the N-region and holes from the P-region are injected into the main active region. When free electrons and free holes coexist with comparable momentum, they will combine and possibly emit photons with energy near the band-gap, resulting in a LED. The process is called injection (or electro-) luminescence, because the injected carriers recombine and emit light through spontaneous emission.

The forward-biased PN junction injects electrons and holes into the GaAs active region. The AlGaAs cladding layers confine the carriers in the active region. A high-speed operation requires high levels of injection (and/or doping) so that the recombination rate of the electrons and holes is very high. This means that the active region should be very thin. However, nonradiating recombination increases at a higher carrier concentration, so there is a trade-off between internal quantum efficiency and speed. Under some conditions, LED performance is improved by using quantum wells or strained layers. For example, a sketch of a LEDs' PN junction is shown in Fig. 2.1. The light emission is produced by the radiating recombination of electrons injected into the P-type material because the electron current is much larger than the hole current. Electrons are injected when the PN junction is forward biased because the injection efficiency relates the current "useful" carriers, which are the electrons injected in the P-type region, to the remaining carriers that are currently in the junction. The junction is where the electron current is injected into the P-type region, and the hole current is injected in the N-type region. Overall, it represents the current of the carriers that are recombining in a nonradiating way. Usually, it reaches a value of 30–60%.

A LED is a simple PN junction made with semiconductor material that exhibits radiating recombination properties. This PN junction can either be a hetero-junction or a homo-junction. The material of the semiconductor's energy band-gap determines the frequency of the emitted light, according to the relationship $E_g = h\nu$. For PN junction LEDs, the wavelength of the emitted light depends on the band-gap of the material. For narrow band-gap materials, the wavelength is in the infrared region, and for wide band-gap materials, the wavelength is in the visible and ultraviolet spectral regions. In addition to the specific band-gap requirements,

Fig. 2.1 Principle diagram of semiconductor light-emitting diodes

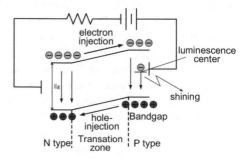

the semiconductor material should be easily doped with both N-type and P-type dopants to form the junction. For example, GaAs is a direct semiconductor material and it can be easily doped with donors or acceptors. GaAs-LEDs usually have a wavelength of around 0.855 μm at room temperature. On the other hand, GaN is another direct semiconductor material, but has a band-gap of ~ 3.40 eV (0.365 μm). A LEDs' wavelength can be tuned easily by choosing ternary materials such as AlGaAs, InGaN, and AlGaN.

2.1.3 Characteristics of a LED

(1) Spectral Characteristics

The spectral characteristics of a LED include the emission wavelength, the spectral width, and the spectral shape. The emission wavelength of a direct-gap LED is determined by the band-gap of the active layer. Because the band-filling effect of both the injected electrons and the holes fills the space near the edges of the conduction and valence bands, this causes the emission wavelength peak to be somewhat shorter than $\lambda_g = h_c/E_g$ and corresponds to a photon energy that is somewhat larger than the band-gap energy. However, if the active layer is heavily doped, the formation of band-tail space can lead to a long emission wavelength, which corresponds to photon energy smaller than the band-gap energy. For an indirect-gap LED doped with isoelectronic impurities, the emission wavelength is longer than λ_g, with photon energy smaller than the band-gap energy.

The emission wavelength peak of a LED varies depending on the injection current and temperature. Because the band-gap of a III–V semiconductor normally decreases with increasing temperature, a LED's emission wavelength peak becomes longer as the operating temperature increases. The rate of change depends on the LED's specific semiconductor material. When the injection current increases, the band-filling effect, which is caused by the increasing concentration of the injected carriers, leads to an increase in the emitted photon energy. This then results in a corresponding reduction to the emission wavelength peak. This effect is often abated by the shrinkage of the band-gap, which is due to the heating of the junction that accompanies the injection current increase.

The spectral width and the shape of the emission are intrinsically defined by the spontaneous emission spectrum. However, a LEDs' emission spectra is often further complicated by frequency-dependent absorption and then scattered by impurities and other materials. These other materials have different band-gaps and are located in the LED's layered structure. In terms of the spectral width of the photon energy, it is approximately $h\Delta v = 3$ k_BT, but it can range between 2 and 4 k_BT. At room temperature, the spectral width of a LED is approximately 80 meV, but it can be as narrow as 50 meV, or as broad as 100 meV, in some devices. In terms of the optical wavelength, the spectral width $\Delta\lambda$ ranges from approximately 20 nm for InGaN-LEDs, which emit short-wavelength ultraviolet or blue light, to the range of

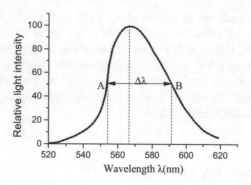

Fig. 2.2   Emission spectrum of a LED

100 nm for InGaAsP-LEDs, which emit long-wavelength infrared emissions. The spectral width of a LED normally increases with both temperature and the injection current. Because a LED emits spontaneous radiation without an optical cavity, the longitudinal and transverse mode structures, which are characteristic of a laser spectrum, do not exist in the LED's emission spectrum. Figure 2.2 shows a representative emission spectrum of a LED.

The spectral characteristics of a LED reflect its strengths and weaknesses of monochromaticity. Many LEDs cannot produce monochromatic light; that is, to say, there is not only one peak wavelength, so we have introduced the term "dominant wavelength" in order to describe the spectral properties of LEDs. A dominant wavelength is defined as the wavelength of main monochromatic light observed by the human eye. A LED has only one dominant wavelength, although some LEDs can give multiple wavelengths of light.

(2) **Light Distribution Characteristics**

Adopting different lens encapsulation structures does not only improve the luminous efficiency of LED, but can also obtain certain light distribution features. Figure 2.3 gives the light distribution characteristics of a common LED.

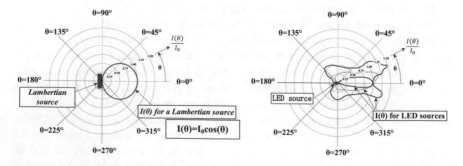

Fig. 2.3   Light distribution characteristics curve of a LED

It is important to note that the LED structure center may not be the optical center. The optical center usually deviates from 5° or more from the structure center. Although there is not much of a problem when measuring with an angle of 40° or more, the deviation should be considered when the view angle is below 40°. Commission International Ed l'eclairage (CIE) recommends that when measuring, it is best to use the structure center rather than the optical center.

(3) **Thermal Characteristics**

The temperature has a great influence on the light output of a LED. The temperature of the PN junction will increase when a current flows through the LED, and this junction temperature change is bound to cause microscopic parameters to change, such as the concentration of internal electrons and holes, forbidden bandwidth, electron mobility. Therefore, corresponding changes take place in the macro-parameters such as the LED light output, light wavelength, the forward voltage. The relationship between the luminous flux output of a LED and the junction temperature can be expressed as:

$$F_V(t_{J2}) = F_V(t_{J1})e^{-K(t_{J2}-t_{J1})} \qquad (2.1)$$

where $F_V(t_{J1})$ is the luminous flux output when the junction temperature is t_{J1}, $F_V(t_{J2})$ is the luminous flux output when the junction temperature is t_{J2}, K is the temperature coefficient associated with luminescent material, and the K-value of AlInGaP and InGaN is about 1×10^{-2} and 1×10^{-3}, respectively.

The diminution of forbidden materials along the bandwidth, combined with the LED's rising junction temperature, will lead to a redshift effect of the LED light wavelength. This relationship can be represented as:

$$\lambda_d(t_{J2}) = \lambda_d(t_{J1}) + \Delta t K_d \qquad (2.2)$$

where $\lambda_d(t_{J2})$ and $\lambda_d(t_{J1})$ are the dominant wavelengths when the junction temperatures are t_{J1} and t_{J2}, respectively; and K_d is the wavelength coefficient that varies with the temperature, which is directly related to the material.

An increase in the LED junction temperature will cause the diminution of the forward turn-on voltage represented by V_F. The relationship can be expressed as:

$$V_F(t_{J2}) = V_F(t_{J1}) + \Delta t K \qquad (2.3)$$

where $V_F(t_{J1})$ and $V_F(t_{J2})$ are the forward pressure drops of a LED when the junction temperatures are t_{J1} and t_{J2}, respectively; and K is the voltage temperature coefficient. The K-value is about -2 mV/°C for main luminescence materials such as AlInGaP and InGaN.

When the junction temperature does not exceed the maximum critical temperature, the forward voltage drop is reversible with temperature change, or else, the LED light output characteristics will be in permanent decline. The highest junction temperature of a LED is related not only to the material, but also to another factors, such as the encapsulation structure. An increase of the LED junction temperature will cause a decrease in the output flux. Once a LED is lit the junction temperature increases along with the consumption of energy, and it only takes a few seconds or minutes for the output to reach thermal equilibrium. Thus, it is necessary to carry out an aging treatment for LEDs before measurement. In addition, with the increase of temperature, the output power of a LED will decrease and a redshift effect will result.

(4) **Current–Voltage Characteristics**

A LED is a kind of PN junction diode. As we have seen earlier, the static current–voltage characteristics of a PN diode can be described by a simple exponential equation:

$$I_F = I_S(e^{\frac{qV_F}{nKT}} - 1) \tag{2.4}$$

where I_S is the reverse saturation current, q is the quantity of electrical charge ($q = 1.602 \times 10^{-19}$ C), n equals 1 or 2, K is the Boltzmann constant ($k = 1.38 \times 10^{-23}$ J/K), and T is the thermodynamic temperature (when the room temperature is 25 °C, $T = 273 + 25 = 98$ K, $\frac{q}{KT} \approx 39/\text{V}$, $\frac{KT}{q} \approx 26\,\text{mV}$).

Like an ordinary rectifier diode, a LED has characteristics such as unidirectional conductivity and nonlinear properties. Figure 2.4 gives the forward volt–ampere characteristic curve of a high-power LED. A forward voltage applied to a LED can produce a forward current. When the voltage is too small, the LED's current is also small and the resistance is large, because the external electric field is not enough to overcome the resistance caused by the internal electric field against the carrier diffusion. But when the voltage exceeds a certain value, the internal electric field is greatly weakened and the resistance becomes very small, so the current increases exponentially. Thus, it is necessary that a precise and constant current source be employed during the process of measurement to retain the stability of the LED light source.

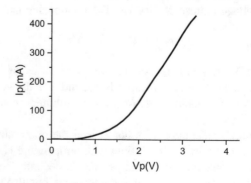

Fig. 2.4 Volt–ampere characteristic curve of a high-power LED

(5) **Output Characteristics**

Inside a LED, the output optical power P_{out} is linearly proportional to the drive current, and this relation defines the output efficiency η:

$$P_{out} = \frac{\eta h v I}{e} \qquad (2.5)$$

This efficiency is strongly affected by the LED's geometry. The internal quantum efficiency, or ratio of emitted photons to incident electrons, is usually close to 100%. The light emitted versus the current produces some nonlinearities, which are much less than a laser diode, but, nevertheless, causes some nonlinearities in the LED's modulation. This LED nonlinearity arises from both the material properties and the device's configuration, and it may be made worse by ohmic heating at high drive currents. The residual nonlinearity is an important characteristic of any LED used in communication systems. Edge emitters are typically less linear because they operate closer to the amplified spontaneous limit.

The performance index of a LED as a light source includes the light and electrical parameters of an electric light source. The electrical parameters consist of the voltage, current, and power. The electrical performance is simple, because the LED is driven by a direct current (DC). However, the current of a LED has no linear relationship with the voltage. Figure 2.5 shows the volt–ampere curve of a LED.

Optical parameters include the luminous flux output, spectral energy distribution, chromaticity coordinates and the color-rendering index, light intensity spatial distribution. The spectral bandwidth range of a monochromatic LED is generally from 20 to 35 nm, which is a narrowband spectrum. The output spectrums of white LEDs vary considerably with the mechanisms of production. The output of the LED light space distribution is directly related to the encapsulation structure. The light angle output of the illuminator LED is usually bigger than the display LED. Figure 2.6 shows the epitome of the LED output characteristic curve.

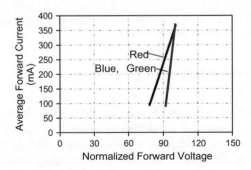

Fig. 2.5 Volt–ampere curve of a LED

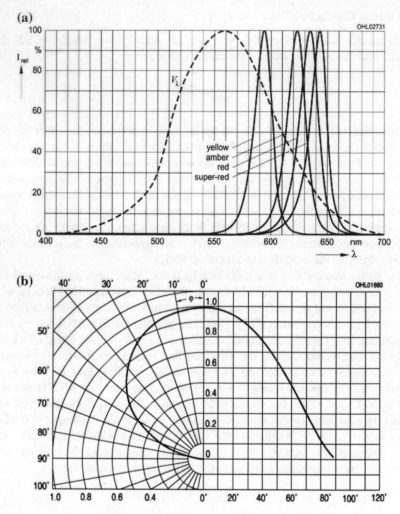

Fig. 2.6 Output performance of a LED based on **a** the spectral energy distribution of a LED and **b** the spatial light intensity distribution of a LED

2.1.4   The Types of White LEDs

The color-rendering index is a unit-less index and is abbreviated variously as either CRI or Ra. CRI is a measure of the degree to which the perceived colors of objects, illuminated by the source, conform to those of the same objects that are illuminated by a reference source for specified conditions. According to the basic principle of white light, we can get the white light from the following several ways.

(1) Complementary colors

By definition, two colors are said to be complementary when their combination yields white light. The corresponding numerical relationship and the luminous power ratio between the wavelengths of complementary monochromatic lights, which are based on the CIE 1964 supplementary standard colorimetric observer, are shown in Table 2.1.

(2) Mixing three colors

White light can be received by combining three colors of light with a particular wavelength, when mixed with a certain power ratio.

(3) Full-wavelength radiation

The entire optical wavelength of light is launched through a broadband generator, which closely resembles the solar spectrum.

(4) Combination of the three methods above

The three former methods can be combined to form white light. For example, the broadband generator and single wavelength generator are used to generate white light.

For LEDs, presently there are three main methods to obtain white light. (1) PC-LED: Yellow phosphor is coated on a blue LED and then a part of the blue light is launched passing through the phosphor with the proper design, which is composed of the blue part of the spectra. At the same time, the remaining portion of the blue light is converted into red and green parts of the spectrum by phosphor. (2) RGB-LED: The white light is produced by adjusting the power of the red, green and blue LEDs all together. (3) UV-LED: White light is produced by coating three kinds of phosphors, such as red, green, and blue, with a UV-LED surface.

Table 2.1 Corresponding relationship between the wavelengths of monochromatic complementary light

Complementary wavelength		Power ratio	Complementary wavelength		Power ratio
λ_1	λ_2	$P(\lambda_2)/P(\lambda_1)$	λ_1	λ_2	$P(\lambda_2)/P(\lambda_1)$
380	60.9	0.000642	460	565.9	1.53
390	560.9	0.00955	470	570.4	1.09
400	561.1	0.0785	475	575.5	0.812
410	561.3	0.356	480	584.6	0.562
420	561.7	0.891	482	591.1	0.482
430	562.2	1.42	484	602.1	0.440
440	562.9	1.79	485	611.3	0.457
450	564.0	1.79	486	629.6	0.668

Note λ_1 and λ_2 are the wavelengths of complementary monochromatic light, and $P(\lambda_2)/P(\lambda_1)$ is the mixed spectral power ratio between complementary colors in order to achieve the same as the D65 chromaticity coordinates

2.2   The PC-LED (Phosphor-Converted LED)

A light source that appears to be white and has a conversion efficiency comparable to that of a fluorescent light source, can be constructed from a blue LED that is covered with a layer of phosphor and then converts a portion of the blue light to yellow light. If the ratio of blue to yellow light is chosen correctly, the resulting light source will appear white to a human observer. This light source is called "LED based on phosphors." In the light source, a part of blue light experiences a Stokes shift after exciting the phosphor and then the phosphor emits a light, which has a longer wavelength than the excitation light. The light color of the light source is determined by the light color of the LED chip and the color of the phosphor coated on the LED chip. By selecting a variety of phosphor colors, a wider light emission spectrum can be obtained and the color-rendering index (CRI) of LEDs can be effectively improved.

2.2.1   The PC-LED's Material and Spectral Characteristics

The YAG: Ce^{3+} phosphor and the $Y_3Al_5Ol_2$: Ce^{3+} phosphor are commonly used in commercial white PC-LEDs because of its broad yellow emission spectrum. After the phosphor absorbs the light emitted by the blue chip, it will emit light from the green band to the red band, which accounts for the major portion of yellow. The yellow spectrum emitted by Ce^{3+}: YAG can be adjusted through the use of other rare earth elements such as terbium and a gadolinium replacing cerium element. Gallium can also be used instead of aluminum in order for YAG to reach the purpose of adjusting its light color. Figure 2.7 gives a typical white light spectrum curve that is obtained from a yellow phosphor excited by a blue LED. Adjustable light parameters can be obtained by adjusting the ratios of blue and yellow light, using different LED main wavelengths, and when YAG phosphors ratios are chosen.

Fig. 2.7 Typical spectrum of blue LED chip + YAG phosphor LED

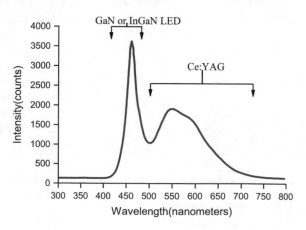

In addition, some sialon materials that are doped with rare earth elements can be used as phosphors, because they also have photoluminescent properties. For example, the element β-SiAlON when doped with the element europium (Europium (II)-doped β-SiAlON) is capable of releasing a strong, wide bandwidth of visible light by absorbing both ultraviolet and visible light. Meanwhile, the polymeric material has a stable crystal structure, and both the luminous brightness and the color will change very little with the change of temperature.

2.2.2   The PC-LED's Structure

The common structure of a PC-LED is shown in Fig. 2.8. The LED chip is fixed in a reflector cup. After the electrodes of the LED are connected to the rear bracket with the gold thread, a mixture of phosphor and silicone is injected into the nearby reflective cup using a traditional type of phosphor coating technology. Through a certain curing process, an external lens made of silica gel or epoxy is used to protect the chip and help the guiding light effect. After the short wavelength of fluorescent light emitted from the LED is absorbed, the higher wavelength light can be emitted.

This way is different from the RGB tricolor white LED, where the light color is entirely dependent on the composition of the LED chip. It is also different from the ultraviolet chip + trichromatic phosphor white LED, where the light color is totally dependent on the composition of the phosphor. It is a compromised and clever way. The white light consists of the light emitted by the LED and the visible light excited by phosphor. The real light flux is a sum of the parts, such as the transmitted blue light and the secondary yellow-emitting phosphor. This method is much simpler and has lower costs. Most high intensity white LEDs on the market are produced using this method. The schematic diagram is shown in Fig. 2.9.

As shown in Fig. 2.10, there are three common distribution methods for the phosphor on the surface of the chip: proximate phosphor distribution, proximate conformal phosphor distribution, and remote phosphor distribution. Remote phosphor distribution requires that the distance between the phosphor and the chip is at least double the lateral length of the chip. Proximate phosphor distribution and

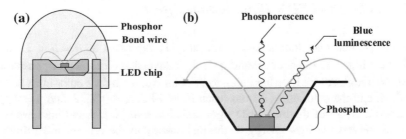

Fig. 2.8 **a** Basic structure of a white LED and **b** wavelength conversion of phosphor and blue luminescence

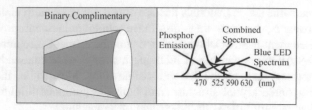

Fig. 2.9 Schematic diagram of a white LED based on blue LED + YAG fluorescence

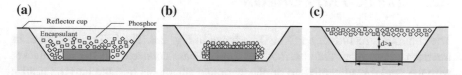

Fig. 2.10 **a** Proximate phosphor distribution, **b** proximate conformal phosphor distribution, and **c** remote phosphor

proximate conformal phosphor distribution mainly focus on improving the uniformity of the white LED color, while the remote phosphor coating method focuses on the promotion of the white LED light output. These three kinds of phosphor coating methods each have their own adherents. Nichia's NS6W083A is produced by using the traditional phosphor coating method. Lumileds uses an electrophoresis coating technology to obtain a uniform white light by coating phosphor on the chip to form a type of thick and uniform coating deposited structure. This technology is also commonly used in Luxeon I and K2 products; however, because the manufacturing cost of the electrophoresis coating process is expensive, the Lumileds Company made a new wrap-type coating method in 2008. The phosphor is directly attached to the chip so that it can convert the blue light into white light. This method has a high degree of color control, and the number of binning can also be effectively reduced.

2.2.3   The PC-LED's Illumination Effect

The second phosphor luminescence efficiency is expressed as $\eta = \eta_v \eta_f K$, where $\eta_v K$ is the blue LED light effect and η_f is the conversion efficiency of phosphor. In order for a high color-rendering index to be obtained, the light's color must be close to white, and the color effect should reach $R_a = 74.7$, $X = 0.3238$, and $y = 0.3264$. Otherwise, a part of the blue LED light radiation will be lost and then cannot be used to convert the light energy to thermal energy in the process of exciting the phosphor to produce yellow. The theoretical limit of the energy conversion efficiency is represented by a ratio, which measures the excitation wavelength and the photon wavelength emission photons. This theoretical maximum conversion efficiency of the phosphor is about $\eta_f \approx 460/560 = 80\%$. Specific data on this matter is shown in Table 2.2.

It is worth mentioning that there are two kinds of light colors to produce white light using both blue LED + YAG phosphors, while the red light is few after 630 nm. The color-rendering index of a LED can be significantly improved by adding a red light component into the LED. However, for high-powered devices, the efficiency of a red LED is lower than a blue LED, so this method will inevitably lead to the decline of final light efficiency.

Making a highly efficient white LED by using conventional phosphor is still a commonly used method. However, in order to manufacture a highly efficient LED through the phosphor method, the largest obstacle to overcome is the Stokes energy loss. In order to achieve a high light output, researchers spare no effort on optimizing the device. For instance, a LED light effect can be improved through enhancing the package design and selecting more suitable phosphors. A conformal coating process developed by Philips Lumileds increases the LEDs' production consistency by overcoming the problem of thickness inconsistency with the traditional fluorescent phosphor coating process. With the development of technology, every new LED produced in the market is accompanied by an increasing luminous efficiency.

2.3   The RGB-LED

The most common method of obtaining white light is by mixing red, green, and blue light; that is to say, by combining monochromatic LEDs, a white light is emitted. This white light source is called the RGB-LED [1]. The most common way to obtain one is by mixing a RGB tricolor light. The schematic diagram of a RGB-LED is shown in Fig. 2.11.

This method is not commonly used in practical applications, because it requires a certain electronic circuitry to control the light color-mixing ratio. However, this method is considered more flexible to obtain the desired light color and high quantum efficiency. In some application areas, this is the first method considered to obtain white light.

The color-rendering index and radiation efficiency of the RGB-LED are affected by the combination of three monochrome LEDs. In order for a high color-rendering index to be obtained, the light color should be close to white, the color effect (R_a) should be over 80, the color coordinates are close to $x = 0.33$, $y = 0.33$, and the ratio of the three-color RGB is set to 1:1.2:1, according to repeated ratio test results. Specific ratio data is shown in Table 2.3.

With the increase in the number of single-color LEDs, the color rendering of the white LED improves, while the light efficiency declines. The expensive price is one of the factors limiting its development, although the tricolor white LED has a high

Table 2.2 Parameters of blue LED + YAG fluorescence

LED	Wavelength (nm)	Spectral bandwidth	Radiation luminosity efficiency (K)	Internal quantum efficiency (%)	Light efficiency (%)	External quantum efficiency (%) = internal quantum efficiency × light out-coupling efficiency
Blue	455	25	36.77	69	50	34.5
Yellow	570	120	420.9		70	
Assumption				80	100	
Theory limitation				100		

LED	Luminous flux	Power (W)	Phosphor conversion efficiency (%)	Secondary emitting blue light needed	Total blue light (W)	Input power	Integrated efficiency (lm/W)
Blue	13.0	0.36			2.31	7.0	65.4
Yellow	420.9	1	51.2	1.95			
Assumption			60		Reach		126.3
Theory limitation			80	1.25			284

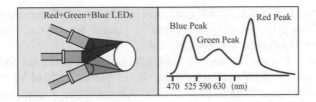

Fig. 2.11 Schematic diagram of a RGB-LED

Table 2.3 Ratios of a RGB tricolor white LED

LED	Wavelength (nm)	Spectral bandwidth	Radiation efficiency	Energy (W)	Flux
Red	614	20	311.6	1	311.6
Green	546	30	640.9	1.2	769.1
Blue	465	20	54.5	1	54.5

luminous efficiency and a good color. In addition, the light fades of a red LED are greater than a blue LED and a green LED. As time goes by, the tricolor white LED will be varying degrees of colors. Therefore, the current RGB white LED is mainly used in LED display fields.

RGB white LED has three types of color combination methods: two-color, three-color, and four-color white LEDs. These three methods behave differently in several areas of light color stability, coloring performance, light efficiency, and other aspects. High light efficiency often means a lower color-rendering index (CRI), so it is truthful to say that you cannot have your cake and eat it too. When the three-color white LED reaches its highest luminous efficiency (120 lm/W), its CRI is the lowest. On the contrary, the four-color white LED often has poor light efficiency despite its excellent CRI.

A multicolor LED provides not only a way to achieve white light, but also a way to achieve different shades. Most colors can be perceived by adjusting the ratio of red, green, and blue hues. However, due to the fact that a LEDs' color is a temperature-sensitive element and will become unstable if the temperature changes during the application, the effect of temperature on LEDs should be emphasized. At a minimum, it should be required that new packaging materials be introduced and an ideal package design is optimized.

2.4   The RGB + UV-LED

A white LED can also be obtained by combining a near-ultraviolet (NUV) LED and a trichromatic phosphor. The trichromatic phosphor is a mixture of three kinds of phosphors, where the red and blue phosphors are based on europium phosphor, and the green phosphor is zinc sulfide (ZnS; Cu, Al)-based and doped with copper and aluminum. The principle of this method is similar to conventional fluorescent lamps, so the department will not repeat it.

The phosphor needed in this method, which falls between 60 and 80 CRI, is readily available, and the visible light part is completely emitted by the phosphor. Therefore, LED has nothing to do with its visible spectrum. A wavelength of about 385 nm UV light, which is emitted by UV-LED chips playing in trichromatic phosphors, can stimulate white light. The conversion efficiency of this part of the phosphor is estimated to be 385/560 = 70%. The theoretical maximum light efficiency of a RGB tricolor white LED is 355 lm/W, while the actual maximum luminous efficiency is 250 lm/W. Figure 2.12 gives the schematic diagram of RGB + UV-LED.

RGB + UV-LED is rare in the practical application because of a relatively large energy loss through the process of converting the UV light into white light, and also because its luminous efficiency is significantly less than that of the RGB white LED. In addition, with the UV-LED chip, it is difficult to obtain deep ultraviolet and the ultraviolet fluorescent wavelength that is emitted is mainly 253.7 nm, so the relevant trichromatic phosphor cannot be moved directly into the LED. As it has been said, it will be a huge challenge to develop a new phosphor suitable for a UV-LED chip. But there is another drawback of the RGB + UV-LED that is often overlooked, which is that when ultraviolet light is leaked from a LED light source, it is harmful toward the eyes and skin of human beings. Considering the luminous efficiency, the application prospect of this approach is not very optimistic Fig. 2.13.

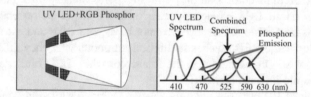

Fig. 2.12 Schematic diagram of a white LED based on UV-LED + RGB phosphor

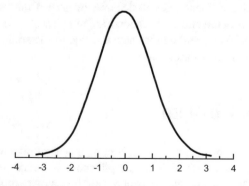

Fig. 2.13 Optical field distribution of a LED

The ultimate aim of comparing these three methods is to achieve the functional lighting of LEDs and further replace traditional lighting. Therefore, the luminous efficiency and color-rendering index are the most concerning parameters regarding LED applications. The developmental direction for the three kinds of white LEDs contains the pursuit of a higher CRI and a higher light efficiency to obtain an excellent color reproduction and a more energy-efficient white LED. Overall, the RGB tricolor white LED can theoretically obtain the highest CRI and light efficiency, and it has a relatively bright prospect with the development of technology.

2.5 The LED's Illumination Light Field and Visual Design

In recent years, LED light technology has been widely used in many public places for energy-saving, environmental protection, uniform lighting, visual clarity, and other purposes. LED lighting technology is also increasingly being used in urban indicator systems through the adjustment of its appearance. In regard to the method of artistic design, LEDs have finally been able to achieve many uses such as a beautiful, practical, and convenient guide with indication actions. In addition, it can also help society achieve the goal of a low-carbon life, while lighting the city's image. However, because of its own structural and luminescence characteristics, LEDs cannot be used for lighting directly and must be designed both appropriately and optically to improve the light distribution of the LED. In this way, they can satisfy the lighting requirements.

2.5.1 Features of the LED Illumination Light Field

The optical design of a LED is very convenient and flexible due to its characteristics of a small size, a near-field approximating surface light, and its far-field approximating point source. Intensive tiling is used to achieve the uniform requirements, which are emphasized in the nonimaging optical design, with a LED as the light source. LEDs are used as backlight LCD displays that use the LEDs' array to achieve uniform light. Compared to traditional light sources, the LED light-emitting angle is not 4π, but 2π, and its light distribution proximity is Lambertian, as shown in Fig. 2.14; therefore, the LED's optical design is very flexible. In some special instances, we can further reduce the LED light-emitting angle and sacrifice some efficiency to achieve special optical design requirements.

The illumination optical system is a nonimaging optical system, and it is different from the conventional imaging optical system, which focuses on the energy distribution rather than the transmission of information. The illumination optical system can be divided into three parts: the light source, the optical system, and the lighting plane. For a vast majority of the illumination optical systems, the requirements of surface lighting are primarily in regard to the light intensity. As a

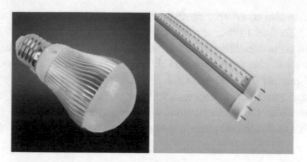

Fig. 2.14 A LED bulb and LED lamp

result, it is necessary that the light intensity distribution, which is produced on the illumination surface, by the light emitted from the light source, through the optical system, be calculated.

The method of calculating this light intensity is called Raytrace. To do so, first the target surface is divided into many small regions with a two-dimensional array. Then, the light that is emitted from different points of the light source transmits to the target surface through the optical system. Each cell in the target surface region can receive a certain amount of light, and then, the illumination distribution of the entire illumination surface can be determined by calculating the number of rays.

2.5.2 The Main LED Optical Design Forms

(1) Direct style: With this way, there is no secondary optical system. The LED light is launched directly without going through an optical device. The optical field distribution can be revised by changing the encapsulated lens of the LED devices so that the light emitted does not follow the Lambertian distribution, but a certain degree of peculiarities contours the map of light.

(2) Diffusion style: In the most lighting conditions, the diffusion optical design is usually used to reduce the surface brightness and expand the luminous range. The typical applications of diffuse types are LED bulbs and LED lamps, as shown in Fig. 2.15.

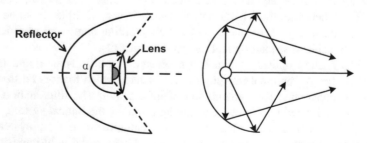

Fig. 2.15 Optical design of the reflective

(3) Reflective style: Fig. 2.16 gives the schematic design diagram of the reflective optical system. The large angle of light is sent out after it is reflected by a surface reflector, and most of the central rays directly emit the light. The disadvantage of this approach is that only a small portion of light efficiency can be controlled, and most of the light is not controlled. Meanwhile, the size of the device is larger, and the position of the LED is not easy to determine and fix. However, different lighting effects can be achieved in practical applications by changing the shape of the reflector and rotating the emitting direction of the LED device.

(4) Transmission type: By adding a lens in front of the LED device, the light's contour map can be changed. The light from the front and back of the lens is shown in Fig. 2.17. Not only are transmissive LED applications more flexible, but there are more diverse designs and it also leads to a lower optical efficiency. In addition, the transmissive type can also control most of the light and it has smaller size.

(5) The combination of reflective and transmissive: As shown in Fig. 2.18, there are two main forms of design. In the first form, the light of the large angle is reflected by the reflector and then emitted through the lens, and finally, the central ray directly irradiates through the lens. This method is flexible, but inefficient, because the light comes into contact with three surfaces. Therefore, the second form can be used to achieve the desired light distribution effect in some cases. In the second form, the lens is moved close to the LED. The center light is emitted after converted by the lens, and then, the reflector reflects the light of the large angle.

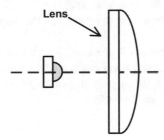

Fig. 2.16 Optical design of the transmissive

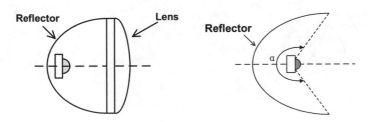

Fig. 2.17 Optical design of the combined reflective and transmissive

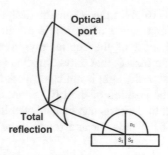

Fig. 2.18   Optical design of total reflection

(6) Total reflection type optical design: The LED light is irradiated to reflectors in the critical angle and then exported by the optical port in a particular direction and particular perspective. The schematic diagram is shown in Fig. 2.19. This way is regarded as a creative optical design, although not commonly used in practical applications.

(7) Second optical design of free-form surfaces: Under normal circumstances, an optical lens is usually used in the optical design of free-form surfaces. Most designs have only a single free-form surface lens, so we must set a surface in the form of a simple curved surface and then design the other surface. A common surface may be a plane, a sphere, or a cylinder. Figure 2.20 shows three common structures used for the free-form surface illumination lens signal:

> ① The inner surface is a sphere, but the outer surface is a free-form surface. A large angle can be achieved, and thus, more applications can be used in the actual design.
>
> ② The inner surface is a plane, but the outer surface is a free-form surface.
>
> ③ The outer surface is a plane, but the inner surface is a free-form surface. Since the outer surface of a plane is restricted, it is difficult to achieve a wide range of light absorption, but it is easy to clean and it has good visual comfort.

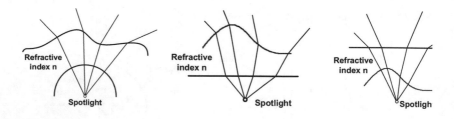

Fig. 2.19   Different types of free-form surface lens designs

2.6 Summary

In recent years, China's LED industry has developed rapidly, and the high growth of markets has also led to advances in technology. Meanwhile, the LED industry has experienced great developments such as the chip, packaging technology, and applications under the strong support of the national and local governments, especially in regard to the strong impetus of the country's fifteen research programs. The advantages of the LED, such as its small size, pure spectrum, and time control are inherent. These characteristics, unlike traditional sources, help the LED own an unlimited numbers of applications. In addition, improvements in optical efficiency, lower prices, and longevity should be achieved from technical progress. In fact, these aspects are the current direction of LED-based technology research.

LEDs, with their inherent characteristics, such as energy saving, life longevity, shock resistance, fast response, cold light source, and other characteristics, take its place in the fields of indicators, blinkers, displays, landscape lighting, and other fields. In recent years, with the pursuit of semiconductor light-emitting materials research, the multicolored and ultra-high bright LEDs have achieved a breakthrough. In chroma aspects, all colors of visible light have been achieved. Most important of all, the emergence of the ultra-high bright white LED makes it possible so that the applications of LEDs can affect the general lighting market.

In summary, LEDs are a new and efficient solid light source and they have many significant advantages such as a long life, fast response, environmental protection, safety, high luminous efficiency, small size, and narrow spectrum, and they are easy to control. The advent of the LED, in theory, will likely be another leap in the history of mankind's lighting creations, next in succession to the incandescent, fluorescent, and high-pressure gas discharge lamps.

2.7 LED Driving

2.7.1 The Physical Device of LED Driving

A visible light transmitter and receiver must be used to achieve interaction information, either in the base station or in the terminal equipment of the VLC system. The typical physical devices used in the experimental system include an optical telescope, a VLC transceiver, an interface driver circuit, a signal processing unit, and a power supply system. Since noncoherent light emitted from the LED light source cannot provide a stable carrier, the current VLC link primarily employs a light intensity (IM) and direct detection (DD) method, where the direct-data connection is used as a communication link. On-off keying (OOK), optical-orthogonal frequency division multiplexing (O-OFDM), and other modulation formats can be used in VLC systems.

The functionality of a LED as a transmitter is based on a fast response time and the modulation of visible light for wireless communications. The dual function of LEDs, such as lighting and communication, causes many new and interesting applications to emerge. A LED is a current-driven unidirectional conducting device, and the brightness is proportional to the forward current. In order to ensure the normal operation of the LED, there are several basic requirements as follows:

(1) The input DC voltage drop may not be less than the forward voltage drop of the LED.
(2) The LED driver circuit should be limited to prevent damage since the high current will shorten the LEDs' life.
(3) There is a certain nonlinear (relationship?) between the LED current and the luminous flux. The current must be controlled into the linear region when we design the VLC system.
(4) The thermal performance of a high-power LED should be noted to prevent damaging the device due to overheating.
(5) The driving circuit should adopt a DC current source or an unidirectional-pulsed current source, rather than a voltage source.

2.7.2   The LED's Driving Mode

According to the load connections, the white LED driver circuit can be divided into a parallel, a serial, or a serial-type hybrid. According to the type of driving source, the white LED driver circuit can be divided either into a voltage-driven or current-driven source. The usual driver classification of a white LED is a combination of the above two categories and includes four kinds of common power-driven approaches, as follows: the voltage source plus ballast resistor, the current source plus ballast resistor, multiple current sources, or the magnetic boost mode, which all drive a series of LEDs. The advantages and disadvantages of these types of power-driven approaches are shown in Table 2.4.

2.7.3   The LED's Drive Circuit Design

A LED made by using a special process produces radiation at forward bias, for example, a PN junction semiconductor device, and the V-I characteristic of a LED is similar to a general diode. However, the LED voltage applied to a PN junction is relatively high. Before the forward voltage reaches the rated value V_f, the current flowing through the LED is small. When it reaches V_f, the current increases very rapidly. Therefore, the LED drive circuit generally employs a constant current

Table 2.4 Comparison of different power supply driving modes

Drive mode	Advantages	Disadvantages
Voltage source plus ballast resistor	Voltage source has a variety of options Just provide a certain voltage Without regard to the forward voltage	Inconsistent LED brightness The higher input voltage, the lower power conversion efficiency Inefficient Inaccurate LED forward current control
Current source plus ballast resistor	Simple circuit Small High current matching performance Low power consumption	Larger power
Multiple current sources	LED current can be adjusted separately without ballast resistors Low power consumption High circuit efficiency Small circuit effective volume	The number of LED drivers is determined by the output port count
Magnetic boost mode drives series LED	Constant current drive without brightness mismatch High power efficiency Great design flexibility and wide application	Inductive element bulky shape High costs EMI-radiated interference

source to output a stable light. For example, limiting an element, such as resistors, must be applied to the direct voltage drive.

A LED generally works at DC power, and a LED drive circuit is usually an AC/DC converter. However, a DC/DC converter can also be used in case there is the presence of DC power. Figure 2.20 gives a simple AC/DC converter circuit. Among them, the input voltage is the strongest and the output constant current is about 20 mA. It can drive 10–16 small powered LEDs. The AC 220 V is rectified to the DC by passing the rectifier bridge, and it then pumps down to about 65 V through the MOS transistor switching circuit. Finally, it outputs a 20 mA current to light the LED by the constant current source. Presently, the power of the LED has reached at least 10 W, and also, the power of a single LED package will become

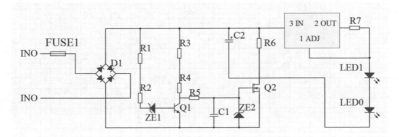

Fig. 2.20 Driving circuit of a GU10 16 low-power LED

increasingly large with the expansion of the LEDs' general lighting. Therefore, high-power LED driver circuits and high-density integrated circuits will be developed.

The LED response time is very short, and the order of magnitude is ns. The light output is substantially proportional to the LEDs' power input or current input. Thus, LED dimming is just as simple as adjusting the current input. In fact, we can also adjust the LED light output by adjusting the duty ratio of the square wave in the way of PWM, which was the square-wave current used to drive the LED.

References

1. Wang, Y., Chi, N.: A high-speed bi-directional visible light communication system based on RGB-LED. China Commun. **11**(3), 40–44 (2014)
2. Bass, M., Stryland, E.: Optical society of America, Fiber optics handbook: fiber, devices, and systems for optical communications. Optical Society of America (2001)
3. Cvijetic, M.: Optical transmission systems engineering. Artech House Publishers (2004)
4. Sze, S.M., NG, K.K.: Physics of semiconductor devices. Wiley-Interscience (2006)
5. Manasreh, O.: Semiconductor heterojunctions and nanostructures. McGraw-Hill Education (2005)
6. JLakowicz, J.R.: Topics in fluorescence spectroscopy. Springer (1991)
7. Narendran, N., Deng, L.: Performance characteristics of lighting emitting diodes. In: Proceedings of the IESNA Annual Conference, pp. 157–164, Illuminating Engineering Society of North America (2002)
8. Craford, G.: LEDs a Challenge for Lighting. LEDs a Challenge for Lighting. Light Sources Lumileds Lighting, LLC, 10 (2004)
9. Tsao, Y.: Solid state lighting: lamps, chips, and materials for tomorrow. IEEE Circuits Devices Mag. **20**(3), 28–37 (2014)
10. Chung, H.Y., Woo, K.Y., Kim, S.J., Kim, T.G.: Improvement of blue InGaN/GaN light-emitting diodes with graded indium composition wells and barriers. Opt. Commun. **331** (22), 282–286 (2014)
11. Den Baars, S.: What is led lighting: technology overview and introduction. DOE SSL Market introduction workshop. Solid-State Lighting and Energy Center (SSLEC), Materials and ECE Departments, University of California, Santa Barbara (UCSB) (2008)
12. Nazarov, P.M.: Luminescence mechanism of highly efficient YAG and TAG phosphors. Moldavian J. Phys. Sci. (2005)
13. Thejokalyani, N., Dhoble, S.J.: Novel approaches for energy efficient solid state lighting by rgb organic light emitting diodes—a review. Renew. Sust. Energ. Rev **32**(5), 448–467 (2014)
14. Tsao, Y.: Solid-state lighting. IEEE Circuits Devices Mag. **20**, 28–37 (2004)

Chapter 3
Models of the Visible Light Channel

For the indoor wireless communication based on white light LED as the new communication system, the establishment of the channel model has not been determined. The measurement and establishment of it are still in its exploratory stages. This chapter makes a preliminary modeling analysis of the visible light channel. The present research on visible light communication systems is mostly based on white LEDs. Thus, first we make a modeling analysis of the generic LED frequency response model, and then secondly, we introduce the physical characteristics and modulation bandwidth of several LEDs, which are commonly used in experiments. Next, we make a summary of the communication links of the indoor VLC. Finally, we make a theoretical and experimental analysis of the photon model of the VLC system from a microscopic point of view, and then thoroughly sum up the channel characteristics of the present VLC system.

3.1 The LED Frequency Response Model

As an important part of the visible light channel, LED frequency response characteristics determine the effective bandwidth of the signal, thereby affecting the transmitting performance of the VLC system. In this section, we make modeling analyzes of the frequency response of both the white and blue components in LEDs, respectively.

3.1.1 The White LED Frequency Response Model

The current commercial white LED products are divided into two main categories according to different spectral components. The first is a blue-light LED chip + the green and yellow phosphor that is used to stimulate the white light, and the second

© Tsinghua University Press, Beijing and Springer-Verlag GmbH Germany 2018
N. Chi, *LED-Based Visible Light Communications*, Signals and Communication Technology, https://doi.org/10.1007/978-3-662-56660-2_3

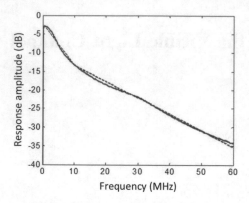

Fig. 3.1 White LED response of the VLC system

is packaging red, green, blue (RGB) three types of LED chips together and mixing them to produce white light. For the first kind of LED, due to the slow response speed of the green and yellow phosphor, the LED modulation bandwidth is very low. The second kind of RGB mixed white LED can provide a high spectral bandwidth, but due to the high cost and complexity of the modulation circuit, it is not widely used in the VLC system design yet. In the following experiments, we mainly use the first type.

The white LED frequency response (dB) that is measured by the pilot signal is shown in Fig. 3.1. There is a significant attenuation of signal high-frequency components. The frequency response curves are divided into two parts to fit linearly. The fitting slope is:

$$s = \begin{cases} -1.02 \text{ dB/MHz}, & 0 \le \omega \le 10 \text{ MHz} \\ -0.42 \text{ dB/MHz}, & 10 \text{ MHz} \le \omega \le 60 \text{ MHz} \end{cases} \qquad (3.1)$$

We then set up a test to see the white LED frequency response using NRZ signals. NRZ signals take samples by four dots per bit, and the spectrum is shown in Fig. 3.2a. After the white LED is modulated, the high-frequency signal is significantly attenuated, and the signal spectrum is shown in Fig. 3.2b.

3.1.2 The LED Frequency Response Model After Blue-Light Filtering

Because the bandwidth of a white LED is very limited, before the signal detection researchers add a piece of blue filter to filter out the yellow light components with a slow response, which increases the modulation bandwidth of the phosphor LED

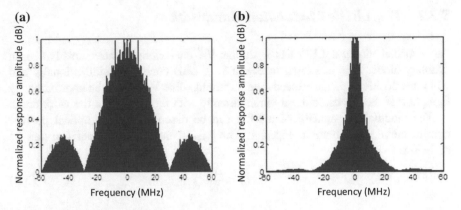

Fig. 3.2 White LED signal spectrum before and after modulation

from 2.5 to 14 MHz. The measured frequency response characteristics of the blue component in this experiment can be expressed as a first-order index function [1]:

$$H(\omega) = e^{-\omega/\omega_b} \tag{3.2}$$

Here, ω_b is the matching coefficient, $\omega_b = 2\pi \times 15.5 \times 10^6$ rads/s.

We fit the frequency response curve (dB), and the fitting slope $s = -0.24$ dB/MHz. The root mean square error between the fitting slope and the actual slope is about $s = 0.08$ dB/MHz.

3.2 The Modulation Bandwidth of Various LEDs

The modulation bandwidth characterizes the modulation capabilities of LEDs. It is an important parameter for the VLC, and an important indicator of a system. A LEDs' response frequency determines the modulation bandwidth of the VLC system, and it is directly related to the data transfer rates. The problem is how to improve the frequency response of the LEDs and how to expand its bandwidth, which must be solved to realize a high-speed VLC. The LEDs' modulation bandwidth is mainly affected by the carrier recombination lifetime in the active region and the PN junction capacitance. In addition to using the LED manufacturing process to reduce the carrier recombination lifetime and the parasitic capacitance, we can also select the appropriate LED chip, which is suitable for communication, by measuring a variety of LED modulation bandwidths. It is necessary to do this because different LEDs on the market have different modulation bandwidths. In addition, we can also use the multi-chip LED due to its great modulation bandwidth potential.

3.2.1   The LED's Modulation Bandwidth

As a special diode, a LED has a similar V-I characteristic curve relative to an ordinary diode. This is shown in Fig. 3.3. A LED conducts unidirectionally and when the positive voltage exceeds the threshold value VA, it can be approximately thought that the electric current varies directly with the voltage in the work area.

The modulation capacity of a LED can be described by the optical power–current curve, as shown in Fig. 3.4. The LEDs' modulation depth (m) can be defined as:

$$m = \frac{\Delta I}{I_0} \tag{3.3}$$

Here, I_0 is the bias current and ΔI is the difference between the peak current and the bias currents. The depth of the optical modulation describes the relationship between the AC signal and DC bias. The greater the depth of the optical modulation, the easier it is to detect the optical signal, thereby reducing the optical power

Fig. 3.3 V-I Characteristic curve of a LED

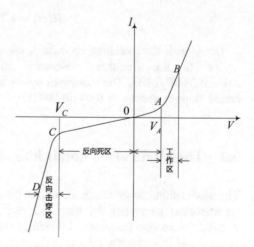

Fig. 3.4 P-I curve of a LED

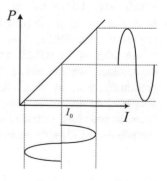

that the receiver requires. The bias current used to drive the LED is often hundreds
of milliamps, so we need to design an appropriate amplifier to make the signal
current reach this magnitude. The drive ability, in most of today's experiments,
achieves a modulation depth of a few percent to more than ten percent. If we only
pursue a high modulation depth, it will lead to a decrease in the modulation
bandwidth, and then have the same effect on the system performance.

The modulation bandwidth of LED determines the transmission rate and channel
capacity of the communication system. It is defined as a frequency when the AC
optical power of the LED drops to half of a low-frequency reference (−3 dB), under
the condition that the degree of modulation remains unchanged. It is shown in
Fig. 3.5. The optical bandwidth in the picture represents the bandwidth when the
signal's current output from the photoelectric detector becomes half of the original.

The modulation bandwidth of a LED is limited by the rate of the response rate,
and the response rate is also affected by the lifetime τ_c of the minority carriers in the
semiconductor:

$$f_{3\,\mathrm{dB}} = \frac{\sqrt{3}}{2\pi\tau_c} \tag{3.4}$$

For the LED made from III-V (e.g., GaAs), τ_c is typically 100 ps, so the theoretical bandwidth of the LED is never higher than 2 GHz. Of course, the total
bandwidth of the current LEDs is lower than this value, and the bandwidth of a
white LED with high power used for lighting is much lower due to the limitation of
microstructure and spectral characteristics. A lower modulation bandwidth limits
the LED's application in the field of high-speed communications, including the
visible light communication system. Therefore, improving the LED's modulation
bandwidth is the key to solve the problem. Literature proposes a method to increase
the modulation bandwidth by improving the LED's microstructure.

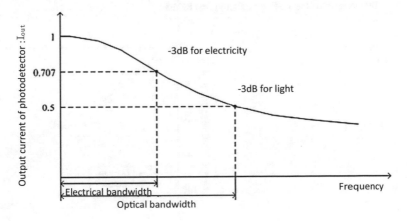

Fig. 3.5 Schematic diagram of a LED's modulation bandwidth

3.2.2 The Modulation Bandwidth of Various LEDs

The equipment used to measure LED modulation characteristics is shown in Fig. 3.6, including both the transmitting and receiving optical signal terminals. To begin, the transmitter amplifies the sine signal generated from the function generator. Then, the amplified signal is loaded into the LED's DC bias current, which is driven by a constant current source. Finally, the LED is able to emit a modulated optical signal, which resembles a flashing light and shade. The receiver, on the other hand, mainly aims at amplifying and processing the photocurrent of the photoelectric detector, and then outputs to the oscilloscope.

Figure 3.7 shows that a 20 dB bandwidth is basically only 25 MHz. A LED modulation bandwidth is primarily limited by its structure. Because various manufacturers use different materials and manufacturing techniques to produce LEDs, there are big differences in the modulation characteristics. The only way to discover the best modulation characteristics is to measure more high-power LEDs. The current commercial high-power white LED is mainly used for lighting. Its internal structure is relatively simple and it does not take the needs of communication systems into account. Some researchers now discuss how to shorten the time of the "rise and fall" in order to improve the modulation bandwidth for high-speed communication systems. They believe the best way to achieve it is by designing a more complex LED microstructure. If one day, we can develop a type of high-power white LED with a wide bandwidth, appropriate luminous efficiency,

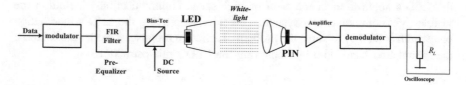

Fig. 3.6 Test system of the LED modulation bandwidth

Fig. 3.7 Modulation bandwidth of different LEDs

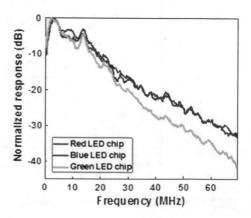

and can be produced massively, it would be the ideal light source of a VLC system. Because the measurement platform of a high-power LED modulation bandwidth is actually a simple VLC system, the modulation bandwidth measurement corresponds to the frequency response bandwidth of the electrical light and electrical channel within the measurement system. Future experiments can try to improve both the electrical path and the optical path between the transmitter and the receiver to compensate for the LED bandwidth. Thus, enhancing the frequency response characteristics of the whole system will increase the transmission rate of the system. On the other hand, we can also equip modulation and demodulation devices based on this experimental platform and create a practical VLC system where we can investigate the performance of the entire communication system.

3.3 Multipath Reflection Modeling

Regarding indoor optical wireless communications, many factors affect the characteristics of the communication channel, such as the communication link pattern, the path loss, and the delay caused by multipath dispersion. These channel characteristics determine the design of communication systems in many aspects, such as the modulation, coding design, power transmitted and the selection of received sensitivity. In addition, we should also consider the influence of the condition parameters to the success of the optical wireless communication system, including the form of the emitted beam, the reception filter, the receiving area and the receiving perspective, among other parameters. All of these parameters are determined by the channel characteristics. Therefore, in order to achieve communication with high reliability and speed, it is absolutely necessary to analyze the characteristics of the indoor optical wireless communication channel.

3.3.1 The Indoor Optical Communication Link Way

Currently, overseas researchers have made a lot of studies relating to the indoor wireless optical communication channel model. Most notably is the indoor wireless communication system channel model studied by the American scholars', Professor John R. Barry, research team [2, 3]. Through their research, they divide the links of the indoor communication system channel into two points. The first point is to see whether the transmitter and receiver are directional. The so-called orientation is actually an angle problem. As for the transmitter, if the divergent angle of the emitted beam is small, and the beam is nearly parallel, we call it a directional transmitter. If the receiver's angular field of view is very small, we call it a directional receiver. If the transmitter and receiver are directional, a link is set up where both the transmitting terminal and the receiving terminal are in alignment, and this link is called a directional link. In contrast, the receiver and transmitter of the non-directional link all have large angles. There is also a link with a mixture of both non-directional and directional characteristics. This hybrid link occurs when one of the transmitters or receivers is directional and the other is non-directional. The second point is whether the undisturbed line of sight (LOS) exists between the transmitter and receiver. In LOS links, the lights received by the receiver not only include the light reflected by other objects, which is emitted from the transmitter with a big angle, but also include the light directly emitted by the transmitter without reflection. However, the non-line of sight (Non-LOS) link is generally the optical signal that the transmitter emits directly toward the ceiling. In addition, the optical signal received by the receiver does not just include the light that comes directly from the transmitter. As a combination of the above two points, the ways to link indoor wireless optical communication systems are divided into the following six methods: a directional LOS link, a non-directional LOS link, a hybrid LOS link, a directional Non-LOS link, a non-directional Non-LOS link and a hybrid Non-LOS link, as shown in Fig. 3.8.

In the indoor VLC system, the LED fixed to the ceiling provides illumination as well as data transmission, so its communication links satisfy two forms of wireless optical communication, both a direct-LOS link and a diffuse link, as shown in Fig. 3.9. In the direct-LOS link, the receiving end and the transmitting end are aligned, so the advantage of these links is high power utilization. However, it requires alignment at both terminals of the link, because once there is an obstacle in the transmission path, the communication will be blocked. Therefore, this is the easiest link scheme, and it is suitable for point-to-point communication as long as

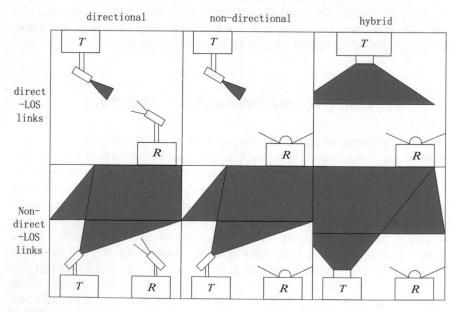

Fig. 3.8 Link modes of indoor wireless optical communication

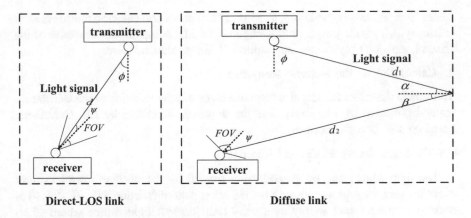

Direct-LOS link   **Diffuse link**

Fig. 3.9 Two kinds of link modes in VLC: direct and diffuse

there are no blocking conditions. Regarding the diffuse link, in order to avoid the system being affected by the shadow effect, we reduce the directional requirements of the transmitting and receiving ends. This allows the receivers to utilize a larger perspective to realize a persistent communication between both the transmitting and receiving ends. The optical power will then be distributed uniformly throughout the chamber, but the multipath effects of the link will limit the signal's transmission rate.

3.3.2 VLC Channel Modeling

1. The VLC Channel Characteristics

Figure 3.10 is a linear baseband transmission model of an indoor VLC system where the impulse response $h(t)$ reflects the channel characteristics.

In a VLC system, the light source of the transmitter is a white LED and the emitted signal modulated by the intensity is $X(t)$. The receiver uses a photoelectric diode for direct detection, and the optical current signal that is received is $Y(t)$, which is expressed as:

$$Y(t) = RX(t) \otimes h(t) + N(t) \tag{3.5}$$

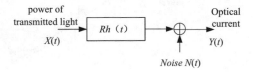

Fig. 3.10 Linear baseband transmission model of an indoor VLC

where $\otimes$ represents convolution, R is the photodiode's photoelectric transformation efficiency, $X(t)$ is the power of transmitted light, $h(t)$ is the impulse response of the channel, and $N(t)$ represents the Additive White Gaussian Noise.

2. Calculation of the Impulse Response

This article describes the impulse response algorithm by referring to the calculation method proposed by J.R. Barry, and the improved algorithm by J.B. Carruthers, based on J.R. Barry's propose.

(1) The Light Source Model and Receiver

The light source can be generally determined by the position vector r_S, unit direction vector n_S, power P_S, and the radiation pattern function $R(\emptyset, \theta)$. $R(\emptyset, \theta)$ is defined as the emitted energy of a solid unit (angle?) light source where $(\emptyset, \theta)$ intersects with n_S. When the radiation source of a transmitter applies the Lambertian model, the radiation intensity of the light source can be expressed as:

$$R(\phi) = \frac{n+1}{2\pi} P_S \cos^n(\phi), \phi \in \left[-\frac{\pi}{2}, \frac{\pi}{2}\right] \tag{3.6}$$

where n is called the Lambertian radiation ordinal and its value is related to the intensity of a half-power angle of the light source. The relationship can be expressed as $n = -\frac{\ln 2}{\ln(\cos \theta_{1/2})}$. Source S can be determined by a triple:

$$S = \{r_S, n_S, n\} \tag{3.7}$$

Similarly, the parameters of receiving unit R include the position vector r_R, the unit direction vector n_R area A_R, and a field of view (FOV). The receiver R can be determined by four tuples:

$$R = \{r_R, n_R, A_R, \text{FOV}\} \tag{3.8}$$

Figure 3.11 shows the direct positional relationship between the light source, the receiver, and the reflector.

Fig. 3.11 Positional relationship between the light source, receiver and reflector

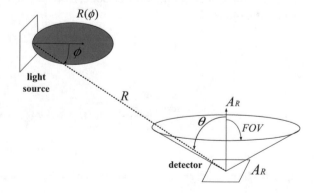

(2) The Reflector Model

We are assuming that all of the emitting surfaces are of ideal Lambertian diffuse, and the radiation pattern is unrelated to the angle of incidence of light. The reflection model occurring on the surface of micro-reflector, which has an area of dA and reflectivity of ρ, is divided into two steps. First, it is considered that the surface of the micro-reflector is a receiver with an area of dA, and we can calculate the optical power dP it receives. Second, we regard the surface of the micro-reflector as an ideal Lambertian source with the power of $P = \rho dP$ and $n = 1$.

(3) The Impulse Response Algorithm

For a specific light source S and receiver R, the impulse response can be expressed as follows:

$$h(t; S.R) = \sum_{k=0}^{\infty} h^{(k)}(t; S, R) \tag{3.9}$$

where $h^{(k)}(t)$ represents the response of the kth reflection.

First, we calculate the impulse response of the zero-order reflection, which represents the light's transfer coefficient when transferring from one point to another without reflected optical power:

$$h^{(0)}(t; S, R) \approx \frac{n+1}{2\pi} \cos^n(\varphi) d\Omega \mathrm{rect}\left(\frac{\theta}{\mathrm{FOV}}\right) \delta\left(t - \frac{R}{c}\right) \tag{3.10}$$

where $d\Omega$ represents the solid angle of the micro-reflector's surface relative to the light source, and it can be expressed by the following formula:

$$d\Omega \approx \frac{\cos(\theta) A_R}{R^2} \tag{3.11}$$

Here, R is the distance between the light source and the receiver:

$$R = \|r_S - r_R\| \tag{3.12}$$

Θ is the angle of incidence of the receiver:

$$\cos(\theta) = \frac{n_R(r_S - r_R)}{R} \tag{3.13}$$

And Φ is the light's included angle relative to the light source axis. Here, the light is emitted from the light source and shined on a receiver:

$$\cos(\Phi) = \frac{n_S(r_R - r_S)}{R} \tag{3.14}$$

The rectangular function is expressed: $\text{rect}(x) = \begin{cases} 1 & for \ |x| \le 1 \\ 0 & for \ |x| > 1 \end{cases}$

The impulse response of the kth reflection can be iterated by the impulse response of the $(k-1)$th reflection through the following formula:

$$h^{(k)}(t; S, R) = \int_S \rho_r h^{(0)}\left(t; S, \left\{r, n, \frac{\pi}{2}, dr^2\right\}\right) \otimes h^{(k-1)}(t; \{r, n, 1\}, R) \tag{3.15}$$

This formula represents the integration of all of the micro-reflectors on S surface. Here, r represents the position vector of the micro-reflector on S surface, n is the normal unit vector of the micro-reflector on surface r, and ρ_r is the reflectance of the micro-reflector on surface r. $R = \|r_S - r_R\|$, $\cos(\Phi) = \frac{n_S(r_R - r_S)}{R}$, $\cos(\Phi) = \frac{n(r_S - r)}{R}$.

3.3.3   A Basic Analysis of the VLC's System Performance

The average transmission power P_t of a light source is defined as:

$$P_t = \lim_{T \to \infty} \frac{1}{2T} \int_{-T}^{T} X(t) \, dt \tag{3.16}$$

The power P_t on the receiving end is defined as:

$$P_r = H(0)P_t \tag{3.17}$$

And here, $H(0)$ is the DC gain of the channel. In the VLC system, the DC gain of the channel is an important characteristic parameter of the direct link channel and it can be obtained by: $H(0) = \int_{-\infty}^{\infty} h(t) \, dt$.

In a VLC system, the quality of received signals can be embodied by a signal-to-noise ratio (SNR). The signal component of the SNR can be expressed as:

$$S = \gamma^2 P_{rSignal}^2 \tag{3.18}$$

where $P_{rSignal} = \int_0^T [h(t) \otimes X(t)] \, dt$ T is the signal cycle. The noise is composed of shot noise, thermal noise, and inter-symbol interference:

$$N = \sigma_{shot}^2 + \sigma_{thermal}^2 + \gamma^2 P_{rISI}^2 \tag{3.19}$$

where

$$P_{rISI} = \int_{T}^{\infty} [h(t) \otimes X(t)] \, dt \tag{3.20}$$

$$\sigma_{shot}^2 = 2q\gamma(P_{rSignal} + P_{rISI})B + 2qI_{bg}I_2B \tag{3.21}$$

$$\sigma_{thermal}^2 = \frac{8\pi kT_k}{G}\eta AI_2B^2 + \frac{16\pi^2 kT_k \Gamma}{g_m}\eta^2 A^2 I_3 B^3 \tag{3.22}$$

q is the electron energy, B is the equivalent noise bandwidth of the receiver circuit, I_{bg} is background current, $I_2 = 0.562$ is the noise bandwidth factor, k is the Boltzmann constant, T_k is the absolute temperature, G is the open loop voltage gain, η is the fixed capacitance of unit area on detector, Γ is the channel noise factor of FET, and g_m is the transconductance of the FET, $I_3 = 0.0868$.

3.4   The Photon Model

3.4.1   The Model Design

1. The Room Model

This section uses a popular room model of 5 m × 5 m × 3 m, where LED lamps are installed on the ceiling and the number and position of the lamps can be set by parameters. The PD array is located on the desktop of at a height of 80 cm and the position on this plane can be set by parameters.

2. The LED Model

We assume that the LEDs are Lambertian point sources S, which can be described by a triple $S = \{\bar{r}_s, \hat{n}_s, m\}$. Where $\bar{r}_s$ is the position vector of the point light sources, $\hat{n}_s$ is the normal direction of the emitting surface where the light intensity is at a maximum, and m is a mode number indicating the light emitting direction:

$$m = -\frac{\ln 2}{\ln \cos \theta_{1/2}} \tag{3.23}$$

In addition, $\theta_{1/2}$ is the included angle between the light, which has ½ of the maximum intensity and $\hat{n}_s$. The light intensity distribution of the light source can be expressed as:

$$Rs(\theta, \phi) = \frac{m+1}{2\pi}\cos^m(\theta), \theta \in \left[0, \frac{\pi}{2}\right] \quad \theta \in [0, 2\pi) \tag{3.24}$$

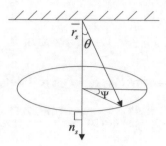

Fig. 3.12 Light source

Here, $Rs(\theta, \phi)$ represents the radiation intensity in the (θ, ϕ) direction. The meaning of (θ, ϕ) is shown in Fig. 3.12.

3. The LED Light Model

A LED light is composed of an array of LEDs. The position, the normal direction, and the modulus of each LED in the array can be set independently.

4. The PD Model

A PD can be represented by four tuples $R = \{\bar{r}_R, \hat{n}_R, A_R, \text{FOV}\}$. Here, $\bar{r}_R$ is the position vector of the PD, $\hat{n}_R$ is the normal direction of the PD's photosensitive surface, A_R is the effective area of the photosensitive surface of the PD, and the FOV represents the incident angle range of the lights that the PD can receive.

5. The PD Array Model

A PD array is composed of multiple PDs. The position, the normal direction, and the FOV of each PD can be set independently.

6. The Reflection Model

The reflective surface uses the Phong Reflection Model. Not only can this model mimic ideal diffuse surfaces, such as walls, ceilings, but it can also accurately model smooth surfaces, such as smooth tile floors, smooth hardwood furniture, and mirror surfaces.

In this model, the incident light is absorbed by the reflecting surface with the probability of $(1 - \rho)$. In addition, the diffuse reflection occurs with the probability of $\rho \cdot r_d$ and the specular reflection occurs with the probability of $\rho(1 - r_d)$. Where ρ is the reflection coefficient of the reflecting surface and r_d is the proportion that diffuses the energy that accounts for the total reflection energy. The spatial distribution of the reflected light intensity is as follows:

$$R_R(\theta_i, \theta_0) = \rho P_i \left[\frac{r_d}{\pi} \cos(\theta_0) + (1 - r_d) \frac{m+1}{2\pi} \cos^m(\theta_0 - \theta_i) \right] \qquad (3.25)$$

Fig. 3.13 Model design of a
receiver and wall reflection

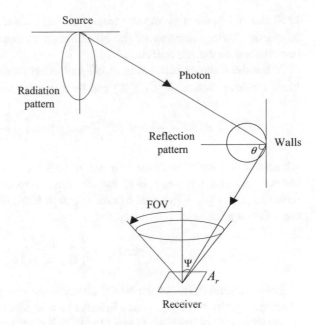

where θ_i is the incidence angle, θ_0 is the observation angle, ρ is the reflection
coefficient, P_i is the power of incident light, and m is the mode parameter which
represents the directivity of the reflected light.

The corresponding light source, receiving and reflecting surface models are
shown in Fig. 3.13.

3.4.2   The Simulation Process and Data Analysis

1. Calculation of the Unit Impulse Response

The unit impulse response of the channel composed of n_1 LEDs and n_2 PDs is the
sum of the unit impulse response of $n_1 \times n_2$ channels composed of a single LED
and a single PD.

The unit impulse response of the channel composed of a single LED and a single
PD can be resolved as follows:

$$h(t; S, R) = h^{(0)}(t; S, R) + \sum_{k=1}^{\infty} h^{(k)}(t; S, R) \qquad (3.26)$$

where S represents the LED, R represents the PD, and the superscript letter k of
h represents the number of reflections that the light reflects from the light source to
the receiver. The first item of the above formula represents the contribution of the

LOS channel to the unit impulse response. The second item of the above formula represents the contribution of the NLOS channel and $h^{(k)}(t; S, R)$ represents the contribution of the kth reflection.

When the distance between the LED and PD is much larger than the size of the PD's sensitive surface, $h^{(0)}(t; S, R)$ can be approximately expressed as:

$$h^{(0)}(t; S, R) \approx \frac{m+1}{2\pi} \cos^m(\theta) \frac{A_R}{d^2} \cos(\psi) \mathrm{rect}\left(\frac{\psi}{\mathrm{FOV}}\right) \delta\left(t - \frac{d}{c}\right) \tag{3.27}$$

where θ is the included angle between $\hat{n}_s$ and $(\bar{r}_R - \bar{r}_S)$, ψ is the included angle between $\hat{n}_R$ and $(\bar{r}_S - \bar{r}_R)$, d is the distance between the light source and the receiver, that is $\|\bar{r}_R - \bar{r}_S\|$, and c is the speed of light. Here, $\mathrm{rect}(x)$ is expressed by the following formula:

$$\mathrm{rect}(x) = \begin{cases} 1 & |x| \leq 1 \\ 0 & |x| > 1 \end{cases} \tag{3.28}$$

For the contributions of the NLOS channel, our group intends to use the Photon Tracing Algorithm (PTA), which belongs to a Monte Carlo method, that we proposed by ourselves to simulate and calculate. It is assumed that there are N photons emitted by the LED (N is very large) and then we track the flight of each photon. Next, we calculate the energy that it contributes to the PD, as well as the time that the contribution occurred. Upon the completion of tracking the N photons, we sum up all of the energy that was received by the PD for a slight time slot on the time axis (e.g., 200 ps) and remove N. Then, we multiply the responsivity of the PD so that we can obtain the unit impulse response during that time slot. All of the points within the two-dimensional plane that use time as the horizontal axis are then connected by a smooth curve, so we can obtain the unit impulse response of the NLOS channel.

The energy of each emitted photon is $1/N$, and the emission direction can be randomly generated according to the following methods.

According to the Lambertian sources' spatial distribution of radiation intensity, it can be obtained that the probability density function of the photons emitted from a LED in the direction of (θ, φ) is:

$$f(\theta, \varphi) = \frac{m+1}{2\pi} \cos^m(\theta) \sin(\theta) \tag{3.29}$$

And the probability distributions of the photons in the direction of θ and φ are respectively:

$$\begin{cases} F_\theta(\theta) = 1 - \cos^{m+1}(\theta) & \theta \in \left[0, \frac{\pi}{2}\right] \\ F_\phi(\phi) = \frac{1}{2\pi} & \phi \in [0, 2\pi) \end{cases} \tag{3.30}$$

Because the occurring directions of the photons can be randomly generated according to the following formula:

$$\begin{cases} \theta = \arccos(\sqrt[m+1]{\alpha}) \\ \varphi = 2\pi\beta \end{cases} \tag{3.31}$$

where α and β are the uniformly distributed random variables on $[0, 1]$.

After being emitted, the photons fly along the straight line simulating the light propagation until they encounter a reflective surface. After colliding with the reflective surface, the photons are reflected or absorbed with a certain probability that is equal to the reflectance. The Phong Reflection Model determines the probability distribution of the flight direction of the reflected photons.

Some photons are reflected with the reflection of the light, which consist of many photons. Only a fraction of the reflected photons can arrive at the PD photosensitive surface to be received and contribute to the unit impulse response. To facilitate the use of the Monte Carlo simulation method, we average the contribution that every reflection of light makes, to the unit impulse response of each reflected photon. Thus, when a photon is reflected, which is equivalent to a point light source with the energy of $1/N$, and N is the total number of photons emitted by the LED, the contribution of the point light source to the unit impulse response is:

$$h_i^{(0)}(t; dA_i, R) \approx \frac{1}{\pi N}\cos(\theta)\frac{A_R}{d^2}\cos(\psi)\mathrm{rect}\left(\frac{\psi}{\mathrm{FOV}}\right)\delta\left(t - T_{PI} - \frac{d}{c}\right) \tag{3.32}$$

Here, i represents the ith reflection of photons, T_{PI} represents the time from when the photons are emitted to when they are reflected, dA_i is an equivalent point light source caused by reflection, and d is the distance between dA_i and R.

Therefore, $h^{(k)}(t; S, R)$ can be expressed as:

$$h^{(k)}(t; S, R) = \sum_{i \in P^{(k)}} h_i^{(0)}(t; dA_i, R) \tag{3.33}$$

where $P^{(k)}$ is the collection of photons not being absorbed after k times reflections.

We then track the paths of all N photons, until either the time that the tracked photons being absorbed, or the flight time, exceeds the time stipulated by the simulation.

After tracking all N photons, we can obtain a number of contributing parameters, which are defined as:

$$C_i^k = \{T_i^k, E_i^k\} = E_i^k\delta(t - T_i^k) \tag{3.34}$$

where C_i^k represents the emitted photons that are being absorbed or the timeout after experiencing k times reflections. When the contribution of the ith reflection to the unit impulse response is C_i^k, T_i^k indicates the happened moment and E_i^k represents the energy contribution to the unit impulse response.

In addition, the time axis is equally divided by interval T_{ts}, and we can calculate the average energy $\tilde{p}_j$ $(j = 0, 1, 2 \ldots)$, which is contributed to by each time slot, respectively.

$$\tilde{p}_j = \frac{1}{T_{ts}} \int\limits_{(j-1)T_{ts}}^{jT_{ts}} \sum_i \sum_k C_i^k(t)\mathrm{d}t = \frac{1}{T_{ts}} \sum_i \sum_k E_i^k \int\limits_{(j-1)T_{ts}}^{jT_{ts}} \delta(t - T_i^k)\mathrm{d}t \qquad (3.35)$$

Here, we multiply $\tilde{p}_j$ by μ (the response of the PD) and draw $(t_j, \mu\tilde{p}_j/N)$ on a two-dimensional plane. We then use a smooth curve to connect them so we can obtain the unit impulse response curve of an indoor wireless optical channel with a single LED and a single PD. Here, t_j represents the moment $\tilde{p}_j$ occurred.

2. **Comparison of the Classical Algorithm and the Photon Tracking Algorithm**

The classic algorithm is mainly used in the simulation of infrared communication systems. Since there is usually only a single light source in the infrared systems, the simulation time of the classical algorithm can be acceptable. In the VLC system, the light source usually uses an LED array, where there are at least one hundred LEDs. If we continue to use the classical algorithm, the simulation time will be too long, so a higher time algorithm is needed. The PTA algorithm is able to meet this requirement.

3.5 Nonlinearity of VLC Communication System

The inherent nonlinearity of LEDs is a challenge for optical orthogonal frequency division multiplexing (O-OFDM). The intensity of the LED can be modulated by a time-varying OFDM signal for wireless access. Since it is modulated by light, a negative number cannot occur when we use light intensity modulation (IM). In order to meet this requirement, it is necessary to modify the OFDM signal to a specific signal format such as DC bias OFDM (DCO-OFDM) and asymmetric clipping OFDM (ACO-OFDM). But this will result in a high peak-to-average power ratio (PAPR) in the transmission of the signal, so the existing VLC system is very sensitive to the nonlinear effects of LEDs.

The LED nonlinear effect affects the OFDM signal by the amplitude distortion at high power and the lower peak clipping at the turn-on voltage (TOV). LED V-I curve shown in Fig. 3.14. LEDs have a minimum threshold value known as the turn-on voltage (TOV) which is the onset of current flow and light emission (below the TOV, the LED is considered in a cut-off region and is not conducting current) Above the TOV, the current flow and light output increases exponentially with voltage (current conduction region). The LED outputs light power that is linear with the drive current.

When the input signal amplitude is too large, LED will work in the nonlinear region and it will show a nonlinear effect. In addition, there is a maximum allowable LED current, the existence of the value will limit the amplitude of the

Fig. 3.14 Nonlinear LED transfer characteristic

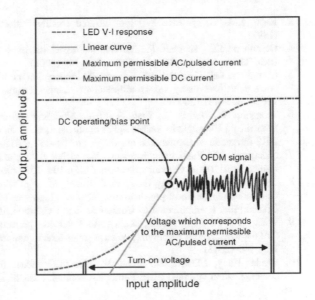

input signal, which makes the OFDM system with high PAPR characteristics can easily show a nonlinear effect. As shown in the figure, LED linear area is not large enough. The carrier density response is related to the frequency, and the frequency response in the passband also introduces the memory effect of the LED [10].

3.6 Summary

This chapter begins with the analysis and modeling of the frequency response of the common LED. It then introduces the physical characteristics and modulation bandwidths of several kinds of LEDs commonly used in experiments. Next, we made a summary of the exposition of the indoor VLC link. Finally, we made a theoretical and experimental analysis on the VLC system's photon model from a microscopic point of view. To finish, we made a comprehensive overview of the present study of the VLC systems' channel characteristics.

References

1. Le Minh, Hoa, O'Brien, Dominic, Faulkner, Grahame, et al.: 100-Mb/s NRZ visible light communications using a postequalized white LED. IEEE Photonics Technol. Lett. **21**(15), 1063–1065 (2009)
2. Barry, J.R., Kahn, J.M.: Simulation of multipath impulse response for indoor wireless optical channels. IEEE J. Sel. Areas Commun. **11**(3), 367–379 (1993)

3. Kahn, J.M., Barry, J.R.: Wireless infrared communications. Proc. IEEE **85**(2), 265–298 (1997)
4. Carruthers, J.B., Kannan, P.: Iterative site-based modeling for wireless infrared channels. IEEE Trans. Antennas Propag. **50**(5), 759–765 (2002)
5. Carruthers, J.B., Carroll, S.M., Kannan, P.: Propagation modelling for indoor optical wireless communications using fast multi-receiver channel estimation. IEEE Proc. Optoelectron **150**(5), 473–481 (2003)
6. Tronghop, D., Hwang, J., Jung, S., et al.: Modeling and analysis of the wireless channel formed by LED angle in visible light communication. In: Information Networking (ICOIN), 2012 International Conference on. IEEE, pp 354–357 (2012)
7. Nakagawa, M.: Fundamental analysis for visible-light communication system using LED Lights. IEEE Trans. Consum. Electron. **50**(1), 100–107 (2004)
8. Zhou, Y., Zhang, J., Wang, C., Zhao, J., Zhang, M., et al.: A novel memoryless power series based adaptive nonlinear pre-distortion scheme in high speed visible light communication. In: Optical Fiber Communications Conference and Exhibition. IEEE, W2A.40 (2017)
9. Wang, Y., Tao, L., Huang, X., et al.: Enhanced performance of a high-speed WDM CAP64 VLC system employing volterra series-based nonlinear equalizer. IEEE Photonics J. **7**(3), 1–7 (2015)
10. Elgala, Hany, Mesleh, R., Haas, H.: An LED model for intensity-modulated optical communication systems. IEEE Photonics Technol. Lett. **22**(11), 835–837 (2010)

Chapter 4
Visible Light Communication Receiving Technology

4.1 The Silicon-Based PIN Photodetector

In the VLC receiving system, photodetectors play a major role. They are a kind of device that transforms optical radiation signals (light energy) into electrical signals (electrical energy). The photodetector is based on the interaction of optical radiation and matter, which is called the photoelectric effect. There are some key indicators of the photodetector such as sensitivity S, responsivity R, the quantum efficiency η, dark current, response rate, wavelength range. In addition to these, we must consider the difficulty and the production process cost, among other factors.

The basic requirements for the photodetector in the visible light communication system are as follows:

(1) A sufficient level of high responsivity in the working wavelength. Namely, to a certain degree of incident optical power, the output photocurrent can be as large as possible.
(2) A sufficient fast response speed, which can be applied to a high-speed or broadband system.
(3) The noise level must be as low as possible to reduce the influence on the signal of the device itself.
(4) A good linear relationship to ensure there is no distortion during the signal conversion process.
(5) Is a small size, has a long service life, etc.

The most commonly used photodetectors in optical communications are PIN photodiodes, avalanche photodiodes (APD), metal–semiconductor–metal photodetectors (MSM-PD), superlattice avalanche photodiodes (SL-APD), waveguide photodetectors (WGPD), and cavity-enhanced photoelectric detectors (RCE-PD), etc.

Among them, the PIN silicon-based plane photoelectric diode has many advantages such as a small junction capacitance, short transit time, high sensitivity,

© Tsinghua University Press, Beijing and Springer-Verlag GmbH Germany 2018
N. Chi, *LED-Based Visible Light Communications*, Signals and Communication Technology, https://doi.org/10.1007/978-3-662-56660-2_4

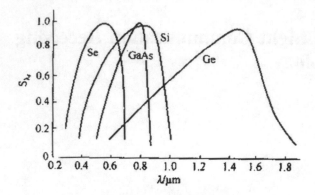

Fig. 4.1 Response curves of various kinds of typical material photodiodes

temperature tolerance, and it is less restricted by applicable occasions. In addition, it has a small volume, is cheap, is lightweight, has good reliability, and is easy to use. Thus, PIN photodiodes have been widely used in optical communications, optical radar, and the rapid photoelectric automatic control field. Figure 4.1 gives the response curves of various kinds of photodiodes made from typical materials. The range of the response wavelength of the silicon-based PIN photodiodes is about 0.4–1.1 um, so it is suitable for visible optical signal detection. In addition, because of the good compatibility with silicon microelectronic technology, the silicon-based PIN photodiodes have become the mainstream of monolithic integrated silicon-based optical receivers. Therefore, most VLC systems adopt these PIN photodiodes.

The core of silicon-based PIN photodiodes is a semiconductor PN junction and the photoelectric effect based on the PN junction. The following work will conclude described details, such as the principle of silicon PIN photodiodes, the characteristic parameters, and the preparation process.

4.1.1   The PIN Structure and Its Working Principle

It is well known that an ordinary photodiode consists of PN junctions, whose structure and formation process is shown in Fig. 4.2. The space charge region is also called the depletion layer, where the neutral zone absorbs most of the incident light. As a result, the depletion layer is usually only a few microns; thus, the PN junction has low photoelectric conversion efficiency, a slow response, and low quantum efficiency.

In order to improve the characteristics of the device, a low-doped intrinsic semiconductor layer is added between the P and N semiconductor material layers to increase the width of the depletion layer w, thus reducing the impact of the diffusive motion and improving the response speed as well. The added layer is called the "I"

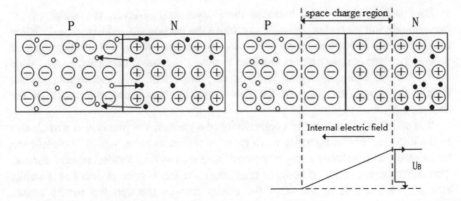

Fig. 4.2 Structure and formation process of the PN junction

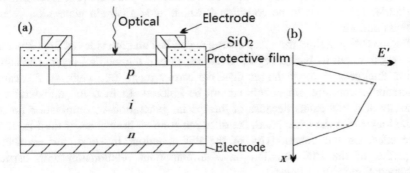

Fig. 4.3 Structure and the electric distribution of a PIN photodiode

layer, due to the low doping density and approximating intrinsic semiconductors. Figure 4.3 shows a PIN photodiode's structure and the electric distribution. It can be found from Fig. 4.3(b) that the I-layer makes up most of the depletion layer; thus, most of the incident light is absorbed in the I-layer and then produces large amounts of electron–hole pairs. On both sides of the I-layer, there are the P and N semiconductors with high doping densities, and they are so thin that they absorb only little incident light. Therefore, the drift component occupies the dominant position in the light current and significantly accelerates the speed of response.

The introduction of the intrinsic layer significantly increases the depletion layer's thickness in p^+ area. This is helpful in order to shorten the diffusion process of the carriers and also significantly reduce the junction capacity, so as to reduce the circuit time constant. In addition, by widening the depletion layer, it also helps to absorb the longwave region. The PIN photodiodes are performing well when the diffusion and drift time is generally in the magnitude of 10^{-10} s and the frequency response is in gigahertz. In the practical application, a major factor that determines the frequency response of a photodiode is the time constant of the circuit, as well as a reasonable choice of load resistance, which is a very important problem as well.

By inserting the I-layer between the P-layer and N-layer, the width of the depletion region increases, in order to achieve the purpose of reducing the diffusion of components. However, an excessive depletion region width will prolong the drift time of the light current carriers and lead to slower response; therefore, a reasonable choice for the depletion region width is very important. This means that we can change the response speed of the PIN diode by controlling the width of the depletion layer.

It is widely known that the conductivity of a general PN junction is mainly due to the diffusion of the minority carriers in the diffusion zone, which is outside the barrier area, and includes a larger forward current as well as a small reverse current. The diffusion area has no electric field area, but the barrier region has a strong electric field, which implies that the carrier transits through the barrier region quickly. That is to say, it ignores the blocking effect of the barrier region. Of course, this process is allowed only in the thin-barrier region. When the PN junction's barrier zone thickness is close to the carrier mean free path, the process of crossing the barrier zone needs to be considered, which adds a certain generation–recombination current.

For the PIN junction, the space charge region lies on both sides of the I-layer and it is very thin, but its barrier region is the made up of the whole I-layer. The barrier region thickness is much larger than the carrier mean free path, so the carrier generation–recombination effects cannot be ignored. In fact, the unilateral conductivity of a PIN exists because of the special generation–recombination process in the I-layer. On the other hand, the diffusion zone on both sides of the I-layer has little effect on the conductivity of the PIN junction. In short, the conductive properties of the PIN junction have an important relationship with carriers' recombination in the I-layer.

When light is incident from the P-region side, the light will extend to the N-region side; meanwhile, the depletion layer absorbs the energy. During this process, the electron is excited from the valence band to the conduction band, and electron–hole pairs are produced, which are namely photogenerated carriers. Through the action of the electric field in the depletion layer, electron–hole pairs drift, respectively, to the N and P-regions and form a light current.

There is no electric field outside the depletion layer region, so the electron–hole pairs generated by the photoelectric effect will disappear when they encounter during the diffusion movement. However, during the process of the diffusion movement, some of the electron–hole pairs with a long diffusion length will extend into the other active layers under the effect of the electric field and the depletion region. Thus, a voltage is generated between the N- and P-layers, which is proportional to the number of separated electrons and holes. If the circuit is connected to the outer, these electrons, through the external circuit, can recombine with holes and form a current.

4.1.2 Parameters

The main characteristics of a PIN photodiode include the following points:

(1) The upper cutoff wavelength:

$$\lambda_c = \frac{hc}{E_g} = \frac{1.24}{E_g} \tag{4.1}$$

For pure Si material, the forbidden bandwidth $E_g = 1.12$ eV, and the upper cutoff wavelength $\lambda_c = 1.1$ μm.

(2) Quantum efficiency η and responsivity ρ:

Photoelectric conversion efficiency can be represented by quantum efficiency η or responsivity ρ. The quantum efficiency is defined as the average number of electrons released from each per incident photon. Assuming the absorption coefficient of the PN junction is $\alpha(\lambda)$, and the junction width is w, then the quantum efficiency is:

$$\eta = \frac{I_p/e}{p/hv} = 1 - \exp(\alpha w) \tag{4.2}$$

To make η bigger, w should be as wide as possible, and that is why the I-layer is introduced in the PN junction.

The responsivity is defined as the current caused by the unit power, namely

$$\rho = \frac{I_p}{P_0} = \frac{\eta e}{hf} \, (\text{A/W}) \tag{4.3}$$

where hf is the photon energy and e is the electron charge.

(1) The quantum efficiency and responsivity depend on the material properties and the device structure. Assuming that the devices' surface reflectivity is zero, the contribution from the N- and P-layers to the quantum efficiency can be ignored, and the I-layer absorbs all of the light in the working voltage, then the quantum efficiency of the PIN photodiode can be approximated as:

$$\eta = 1 - \exp[-\alpha(\lambda)w] \tag{4.4}$$

where $\alpha(\lambda)$ and w represent the absorption coefficient and thickness of the I-layer, respectively. As we can see from the above equation, when the $\alpha(\lambda)$ $w \gg 1$, $\eta \to 1$, so in order to improve the quantum efficiency η, thickness of the I-layer w should be sufficiently large.

(2) The spectral characteristics of the quantum efficiency depend on the absorption
spectrum of the semiconductor materials $\alpha(\lambda)$. The long-wavelength limitation
is determined by the upper cutoff wavelength, namely $\lambda_c = hc/E_g$.

Figure 4.4 shows the spectral characteristics of quantum efficiency η and
responsivity ρ. It can be seen from Fig. 4.4 that Si is best suited for the 0.5–0.9 μm
band, both Ge and InGaAs are best suited for the 1.3–1.6 μm band, and the
responsivity is generally 0.5–0.6 (A/W). You can see that the responsivity and
quantum efficiency vary with the wavelength. In order to improve the quantum
efficiency, we need to reduce the reflectivity of the incident surface so that more
incident photons can move into the PN junction, thus increasing the width of the
depletion region so that the photons will be fully absorbed in the depletion region.

(3) The bandwidth is represented as:

$$BW = \frac{1}{2\pi(\tau_{tr} + \tau_{RC})}, \quad \tau_{tr} = \frac{w}{v_{sat}}, \quad \tau_{RC} = R_s C_d \tag{4.5}$$

where τ_{tr} and τ_{RC} show the transmission time and discharge time, respectively, and
v_{sat} is the carrier's drift velocity. The parameter of BW is very important in optical
communication.

(4) The volt–ampere characteristic curve

Figure 4.5 shows the volt–ampere characteristic curve of a PIN photodiode, which
works in the third quadrant, namely in photoconductive mode. The current generated without light is called the dark current. Typically, the Si photodiode has a
dark current minimum, and the Ge-PIN photodiode has a dark current maximum,
which is one of the main reasons that the Si-PIN photodiode is favored more.

(5) The response speed

The transmit time and equivalent circuit of the PIN photodiode limit the response
speed of the PIN photodiode, as shown in Fig. 4.6.

Fig. 4.4 Quantum efficiency of several typical material photodiodes

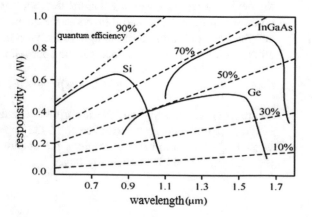

Fig. 4.5 Volt–ampere
characteristic curve of a PIN
photodiode

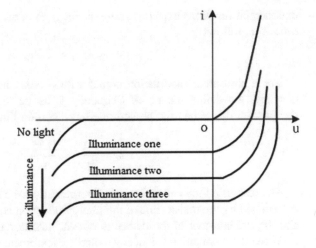

Fig. 4.6 Equivalent circuit of
the PIN photodiode

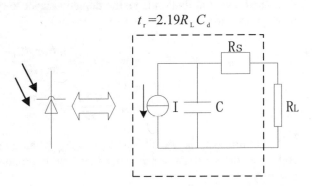

The rise time t_r represents the time that the output current rises from 10 to 90%
of the maximum.

$$t_r = 2.19 R_L C_d \tag{4.6}$$

(6) The response time and frequency characteristics

We can use the pulse response time τ or the cutoff frequency f_c to show the response
speed of the photodiode to the high-speed modulated optical signal. For the digital
pulse modulation signal, from the time that the leading edge of the light-current
pulse rises to 90% from 10% of the maximum, and the lagging edge falls from 90 to
10%, it can be, respectively, defined as the pulse rise time τ_r and the pulse fall
time τ_f.

When the photodiode has a certain time constant τ_0, the pulse has the same pulse
leading and lagging edge, and the pulse edge is, respectively, close to the

exponential functions $\exp(t/\tau_0)$ and $\exp(-t/\tau_0)$, then the resulting impulse response time is as follows:

$$\tau = \tau_r = \tau_f = 2.2\tau_0 \tag{4.7}$$

For a sinusoidal modulation signal with a constant amplitude and frequency $\omega = 2\pi f$, we define the cutoff frequency f_c as the light-generated current $I(\omega)$ decreases 3 dB. When the photodiode has a certain time constant τ_0:

$$f_c = \frac{1}{2\pi\tau_0} = \frac{0.35}{\tau_r} \tag{4.8}$$

The response time and frequency characteristics of a PIN photodiode are mainly determined by the transit time of the photoproduction carriers in the depletion layer and the RC constant of the detection circuit including the photodiodes.

When the modulation frequency ω and the reciprocal of the transit time τ_d can be compared, the contribution from the depletion layer to quantum efficiency $\eta(\omega)$ can be expressed as:

$$\eta(w) = \eta(0)\frac{\sin(w\tau_d/2)}{w\tau_d/2} \tag{4.9}$$

$$f_c = \frac{0.42}{\tau_d} = 0.42\frac{v_s}{w} \tag{4.10}$$

In the formula, the transit time $\tau_d = w/v_s$, where w is the depletion layer width and v_s is the carrier's transit speed, which is proportional to the electric field intensity.

As seen from the above equation, we can reduce the transit time τ_d by decreasing the width of the depletion layer w, thereby enhancing the cutoff frequency f_c. However, it will reduce the quantum efficiency η at the same time.

The cutoff frequency is limited by the circuit's RC time constant:

$$f_c = \frac{1}{2\pi R_t C_d} \tag{4.11}$$

In the formula, R_t is the total sum of the series resistance and the load resistance of the photodiode. In addition, C_d is the sum of the junction capacitance C_j and the distributed capacitance, and $C_j = \frac{\varepsilon A}{W}$, where ε is the dielectric constant, A is the junction area, and w is the depletion layer width.

(7) Noise

The noise will affect the sensitivity of the optical receiving system. The noise of a PIN photodiode includes shot noise i_s, the thermal noise i_t, and the flicker noise $\frac{1}{f}$. Shot noise originates from the signal current and the dark current, and thermal noise is generated by the load resistance and the input resistance of the amplifier.

The standard root-mean-square of shot noise: $i_s = \sqrt{2eI_p}$, where I_p is light current.

The standard root-mean-square of dark current noise: $i_d = \sqrt{2eI_\alpha}$.

The standard root-mean-square of thermal noise: $i_t = \sqrt{\frac{4kTB}{R}}$, where $k = 1.38 \times 10^{-23}$ J/K is the Boltzmann's constant, T is the equivalent noise temperature, and R is the parallel result of the load resistor and the amplifier input resistor.

Flicker noise $\frac{1}{f}$ can be neglected at high frequencies.

In a visible light communication system, not all of the photogenerated carriers are signals, and only carriers generated between a 420 and 480 nm visible light wavelength are required signals of the optical communication system. Carriers generated by near infrared (from 400 to 1100 nm), and other outside communication wavelengths, are also classified as noise. However, the response peak of a Si-based PIN is usually between 800 and 900 nm, which is much higher than the communication 420–480 nm band in terms of efficiency. So if no other measures are used, the communication signals will be drowned by strong background noise caused by disturbing light.

(8) The signal-to-noise ratio

The signal-to-noise ratio (S/N) of a PIN photodiode is defined as:

$$S/N = \frac{I_p^2}{2e(I_p + I_d)B} + \frac{4k_B TB}{R_e} \tag{4.12}$$

where I_p is the peak current, I_d is the dark current, B is the bandwidth, R_e is the equivalent resistance, and I_p is defined as:

$$I_p = e\eta \frac{P_{opt}}{\hbar\omega} \tag{4.13}$$

Here, η is the quantum efficiency, which is the photoelectron acceptation rate from both ends, e is the electronic charge, and P_{opt} is the absorbed light power. The working characteristic parameters of different material PINs are shown in Table 4.1.

Table 4.1 Working characteristic parameters of different material PINs

Parameters	Symbol	Unit	Si	Ge	InGaAs
Wavelength range	λ	nm	400–1000	800–1650	1100–1700
Responsivity	R	A/W	0.4–0.6	0.4–0.5	0.75–0.95
Dark current	In	nA	1–10	50–500	0.5–2.0
Rise time	τ	Ns	0.5–1.0	0.1–0.5	0.05–0.5
Bandwidth	B	GHz	0.3–0.7	0.5–3.0	1.0–2.0
Bias	VB	V	5	5–10	5

4.1.3   The Device Preparation Technology

As mentioned above, on both sides of the PIN diode there is a heavily doped n^+ and p^+ semiconductor and an intrinsic semiconductor, the I-layer, with a high resistivity inserted between them. In fact, because of the material and process, the middle layer cannot be the ideal intrinsic semiconductor, and it means that the real "I" does not exist. It must contain a small amount of impurities, and the typical value of the resistivity is about 1000 Ω cm. If the I-layer contains a P-type impurity, we call it the P-layer, and if it contains N-type impurities, we call it the N-layer, so the actual PIN diode is actually a $p^+\Pi n^+$ tube or a p^+vn^+ tube (Fig. 4.7).

There are three main methods to fabricate the Si-based PIN diode: an ion implantation method, a diffusion method, and a bonding method, which are introduced as follows:

(1) The Ion Implantation Method

N-type silicon material's resistivity is 1500 to 5000 Ω cm, thickness is 140 µm, and the injection power can be 150 keV. Next, inject boron 1×10^{16} cm^{-3} in one side of the silicon, and on the other side, inject phosphorus 1×10^{16} cm^{-3}; thus, it forms the n^+ and p^+. The final result of the experiment shows that the reverse breakdown voltage is 1000 V and the depletion region capacitance is 3 PF, when the diode diameter is 60 um and the carrier lifetime is 15 us. Ion implantation is better than using diffusion because it can maintain a better interface uniformity.

(2) The Diffusion Method

N-type silicon material's resistivity is 1500 to 5000 Ω cm, and the thickness is 140 µm. When using the trebling diffusion method in the N-type silicon, boron and phosphorus are used to diffuse both sides, respectively. The spread depth of the p^+-n junction is 23 µm, and depth of the n^+-n junction is 20 µm. The reverse breakdown voltage is 1000 V, and the depletion region capacitance is 3 PF. When the diameter of the diode is 60 um, the carrier lifetime is 15 us. When the layered distribution is this kind of diode, it is a multilayered diode structure.

Fig. 4.7 A PIN diode structure

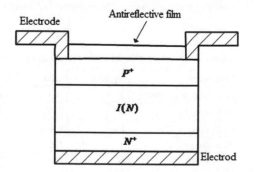

(3) The Bonding Method

Silicon wafer bonding technology can bond together silicon and silicon, silicon and glass, and other materials through both chemical and physical effects, including static bonding and hot bonding. Static bonding can bond metal and glass, metal and a semiconductor without any adhesive and offers a bonding interface with good air tightness and long-term stability. Hot bonding bonds two wafers via high-temperature treatment instead of using any adhesive or electric field. The bonding method has become a research hot spot because of its lower interface defects and low manufacturing cost advantages.

Factors that influence the device performance during technology preparation mainly include the following:

(1) Choosing high quality, high resistance materials that can be depleted completely under a low bias voltage. Also, a small device capacitance is helpful in order to reduce noise and improve the response speed.
(2) Slow annealing in low temperatures and the chip phosphorus absorption process, which can keep perfect lattice and reduce the dark current.
(3) Low-energy ion implantation to realize shallow junctions, reduce the "dead zone," and improve the short-wavelength quantum efficiency.
(4) The deposition of a reasonable thickness of silicon nitride/silica, which coats the sensitive area window and will enhance the spectral responsivity of the short wavelengths.

4.2 The Narrowband Blue Light Detector

A light detector is very important in the visible light communication system, because it converts light signals into electrical signals. A conventional photodetector is a device with a wide spectral response, which can respond with photons whose energy is greater than its forbidden bandwidth. A photoelectric detector shows a sharp cutoff characteristic at a long wavelength. There is a simple relationship $\lambda_c = 1.24/E_g$ (μm) between the cutoff wavelength λ_c and the forbidden bandwidth materials E_g (eV). When approaching the cutoff wavelength, the response falls sharply, and then there is not any response. On the short-wavelength side, the cutoff characteristic is a slow decline, and the decline rate varies according to the material and the devices' design. However, it generally does not have an obvious wavelength cutoff. The response spectrum of the photoelectric detector can be cut in a certain range to adapt to specific requirements through the reasonable design.

A variety of photoelectric detectors can be used in a visible light communication receiver, and the silicon photoelectric detector is undoubtedly the first to be considered (see Sect. 4.1 of this chapter) because of its mature technology, good photoelectric performance, and low price. The forbidden bandwidth of monocrystalline silicon materials is 1.12 eV at room temperature, so the response of the silicon photoelectric detector extends to more than 700 nm, which is a near-infrared wave band, and the cutoff wavelength is about 1.1 μm. A silicon photoelectric detector can have a good response throughout the entire visible spectrum with reasonable materials and a good device structure design, which means that the response on the shortwave end can also be extended into the ultraviolet band. As the optical channel has the characteristic of a narrowband in visible light communication, the light wavelength used during communication is generally within a narrow range. For example, a LED blue light in a fluorescent LED, or other monochrome LED, or even a laser, is generally used in a narrow range. So, in principle, the silicon photoelectric detector itself is not the best technological choice because of its wide response spectrum. For visible light communication, an optical channel is set up in the free space, so of course we hope the photoelectric detector responds to the light wavelength channel as fast as possible, and the wavelength out of the light channel can be completely avoided. Free-space interference includes natural light (the sun), yellow light weight in fluorescents, as well as other various light sources. Light interference is usually uncontrolled regarding the visible light communication system. A photoelectric detector will produce a strong noise, some burst blocking, a saturated device, and a series of adverse effects, if the photodetector can respond to light of different wavelengths. To avoid these impacts, a limitation of the response spectrum width to the photoelectric detector is required for the visible light communication system. In general, the interference light will resemble a long wavelength in terms of visible and near-infrared light. In addition, the response of the conventional silicon photodetector is signaled with long wavelengths, whose peak response wavelength is above 900 nm. The long-wavelength band's peak degree response is twice as high as in the blue light band, so the impact of the long wavelength will be greater. For the silicon photoelectric detector, the use of a filter, such as a longwave-cutoff filter or a narrowband-pass filter. (see Sect. 4.3 of this chapter), is a very effective way to limit the spectral range response. However, the introduction of a filter will reduce the response of the device, increase the complexity of the device and the cost, and reduce the reliability as well. Besides, the design and production of a high-performance narrowband filter has its own certain difficulties. So the photoelectric detector, which matches with narrow channel wavelength's response spectrum characteristics, is of course the desired technology for the visible light communication system. And, of course, other parameters such as quantum efficiency, dark current and noise characteristics, response speed of the photoelectric detector also need to satisfy the system requirements. In this section, we will show the general principles used when choosing the parameters of a narrowband photoelectric detector in terms of the blue channel. In addition, we will introduce the InAlP narrowband blue light detector devices, which are suitable for the application.

In Fig. 4.8, the solid line represents the direct band-gap area, the dotted line represents the indirect band-gap area, and the arrow represents the In composition matching the GaAs substrate. When we use an LED light as the channel source of the visible light communication system, the spectrum width of the emission light is

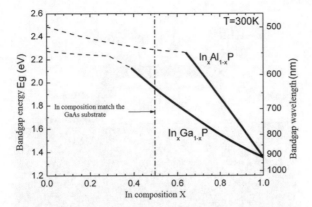

Fig. 4.8 Relationship between the forbidden bandwidth E_g, the band-gap wavelength, and the In material components of the III–V semiconductor materials $In_xAl_{1-x}P$ and $In_xGa_{1-x}P$

about 10% of the center wavelength and is a symmetric Gaussian distribution on both sides of the center wavelength. For the blue LED, its center wavelength is 480 nm and the spectral width is usually about 50 nm, namely 480 ± 25 nm. In addition, it is the same case with that of blue light weight in fluorescence white LEDs as well. For the blue channel, the cutoff wavelength of the photoelectric detector in a long wave must be greater than 480 nm. Commonly, 500 nm is more appropriate, which corresponds to the photon energy of about 2.5 eV. In general, the performance of the wide forbidden band material's photoelectric detector will be better, especially if the dark current is lower and both the temperature and stability characteristics are better. Therefore, when we design such a photoelectric detector, the forbidden bandwidth materials should be close to, but slightly less than, 2.5 eV. Figure 4.8 shows the relationship between the forbidden bandwidth E_g and the band-gap wavelength of the III–V semiconductor materials $In_xAl_{1-x}P$ and $In_xGa_{1-x}P$ with In material components. With the components of $InxAl_{1-x}P$ and $In_xGa_{1-x}P$, 0.48 and 0.49 apply, respectively, under the conditions that the GaAs substrate matches the lattice. The corresponding band-gap is 2.3 eV (indirect band-gap) and 1.9 eV (direct band-gap). Therefore, III–V $In_{0.48}Al_{0.52}P$ is very suitable for the blue-ray channel of visible light communication as an absorption layer of the photoelectric detector. In addition, the band-gap wavelength is 540 nm, which happens to meet the demands of the longwave side. In this material system, the III–V $In_{0.49}Ga_{0.51}P$ material has a smaller band-gap than $In_{0.48}Al_{0.52}P$, but $In_{0.49}Ga_{0.51}P$ can match the GaAs substrate. What's more, $In_{0.49}Ga_{0.51}P$ is the direct band-gap that has a higher light absorption coefficient, which means $In_{0.49}Ga_{0.51}P$ can be a window-layer photodetector to limit response at the shortwave side, and achieve the goal of cutting a shortwave side carrier. Before making the actual devices, we need to perform in-depth research and analyze the growth process, material structure, and the optical and electrical properties of $In_{0.49}Ga_{0.51}P$ and $In_{0.48}Al_{0.52}P$. Afterward, we can design the device structure on the basis of our findings.

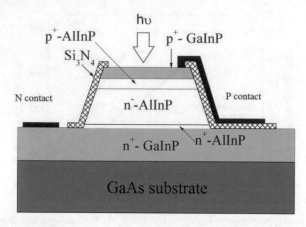

Fig. 4.9 Epitaxial material and device structure diagram of the $In_{0.48}Al_{0.52}P/In_{0.49}Ga_{0.51}P$ narrowband blue light detector

Figure 4.9 shows the epitaxial material and device structure diagram of the $In_{0.48}Al_{0.52}P/In_{0.49}Ga_{0.51}P$ narrowband blue light detector. The detector material is grown through the method of gas source molecular beam epitaxy in GaAs substrate. A 1-μm high-doped N-type $In_{0.49}Ga_{0.51}P$ buffer layer is grown first, which is also called the N-type ohm contact layer. Then, 50 nm of high-doped N-type $In_{0.48}Al_{0.52}P$ is grown as a back surface field (BSF) layer, and 1 μm thick of low-doping N-type $In_{0.48}Al_{0.52}P$ is grown as a light absorption layer. A 150-nm high-doped P-type $In_{0.48}Al_{0.52}P$ is then added to form a PN junction and covers the 150-nm high-doped P-type $In_{0.49}Ga_{0.51}P$ as a shortwave limiting layer, which is also called the P-type ohm contact layer. After these processes, we can encapsulate the detector with a conventional compound semiconductor encapsulation device.

Figure 4.10 shows the optical spectral response characteristics of the $In_{0.48}Al_{0.52}P/In_{0.49}Ga_{0.51}P$ narrowband blue light detector. We can see that there is a

Fig. 4.10 Optical spectral response characteristics of the $In_{0.48}Al_{0.52}P/In_{0.49}Ga_{0.51}P$ narrowband blue light detector

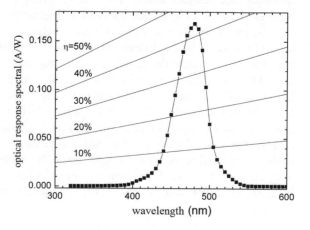

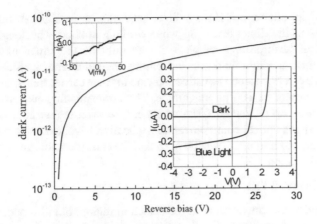

Fig. 4.11 I-V characteristics of a detector in room temperature

response peak at 480 nm, and the spectral width is about 10%, which would match well with the emission spectrum of a blue LED. By adjusting the absorption layer of $In_{0.48}Al_{0.52}P$, and the shortwave limiting layer thickness of $In_{0.49}Ga_{0.51}P$, we can fine-tune the peak response wavelength of the detector and the cutoff characteristics of both the long wave and shortwave. Once the detector's external quantum efficiency on the peak wavelength reaches 43.4%, we can continue increasing the external quantum efficiency by adding reflector-resistant coating. The detector has a very low dark current because of the wide band-gap materials.

Figure 4.11 shows the I-V characteristics of the detector in room temperature, while the diameter of photosensitive surface is 300 μm. We can see from the figure that the dark current is about 3 pA, when the detector's negative bias is commonly 2–3 V. The product of zero resistance with the area, namely R_0A, can reach 6.9×10^8 Ω cm^2 at room temperature and has higher output voltage than a silicon detector when working in photovoltaic mode.

The response speed is another factor to be considered in the photoelectric detector's communication system. Three aspects generally affect the photoelectric detector's response speed:

(1) The Optical Carrier Transit Time

For a reasonably designed device, light absorption primarily occurs in the depletion region, so the transit time directly depends on the width of the depletion region and the carrier drift velocity. Due to the high electric field strength in the depletion region, the carrier drift velocity can be close to the saturation speed v_s, and the conventional v_s of a semiconductor can reach more than 10^6 cm/s. In order to produce effective light absorption, when the light absorption coefficient is α at the detecting wavelength, we hope that the thickness of the absorbing layer can reach $1/\alpha$ to ensure enough quantum efficiency. Similar to the ternary system of the compound semiconductor like $In_{0.48}Al_{0.52}P$, its band-gap energy is very close to the direct band-gap, although it is an indirect band-gap, so its light absorption coefficient is close to the direct band-gap when working at short wavelengths. In addition, the light absorption coefficient is generally at the order of a 10^4 cm^{-1} magnitude; thus, a 1 mm thickness can ensure enough quantum efficiency. In this condition, we can have a conservative estimate that the carrier transit time $t(t \propto 1/\alpha v_s)$ is at the orders of 100 ps magnitude. The corresponding response frequency $f = 1/2\pi t$ can be up to several GHz, which can be even higher after optimization. For the free-space optical communication applications, we generally require a code speed in dozens to hundreds of Mbit/s, so that the carrier transit time does not need special consideration.

(2) The Diffusion Effect

Light absorption not only occurs in the depletion region, but also occurs in its two sides. Therefore, the photoproduction carrier outside of the depletion region needs to reach the depletion region by the diffusion function so as to produce an effective photocurrent. Compared to the drift, diffusion is a slow process, so the diffusion function results in producing a trailing impulse response for the photodetector. For the photodetector, the tail phenomenon always exists, but we can reduce the impact effectively through a reasonable device structure design and by imposing a certain reverse bias on the detector. We can eventually eliminate its influence through a subsequent simulation balanced sentence and digital circuit. As for the narrowband blue light detector we designed, due to the design of a thinner 150-nm high-doped P-type $In_{0.48}Al_{0.52}P$ layer, the diffusion effect is completely reduced.

(3) The Influence of the RC Time Constant

Because the photodetector has a certain capacity, including the junction capacitance and distributed capacitance, and both the series resistance and preamplifier circuit have a certain impedance directly connected with the photodetector in the whole optical receiver, its overall RC time constant may ultimately determine the system's response speed. The capacitance of the photodetector is not only proportional to the device's photosensitive area, but it is also influenced by the dielectric material properties, device structure, and the influence of the reverse bias.

Figure 4.12 shows the C-V characteristics of the $In_{0.48}Al_{0.52}P/In_{0.49}Ga_{0.51}P$ narrowband blu-ray detector when the photosensitive surface diameter is 300 mm. Even under the zero bias, the narrowband blue light detector's light absorption layer is in a state of completely running out due to its active light absorption layer's low doping concentration. Also, the series resistance is so small that you can ignore its effects, thanks to the high-doping contact layer device. We can see from the figure, if a 3 V reverse bias is applied, the capacitance is about 8.7 pF. When a high-frequency amplifier is used, which has an input impedance of conventional 50 Ω, the frequency response can reach more than 350 MHz, which is determined by the RC time constant. For practical application, a more convenient method is to use a transresistance preamplifier to form a complete set of limiting amplifier/clock recovery circuits, which is very mature in optical fiber communication technology.

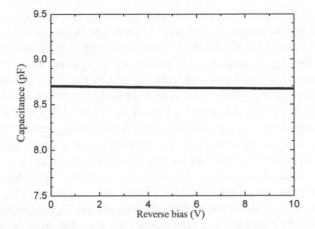

Fig. 4.12 C-V characteristics of the $In_{0.48}Al_{0.52}P/In_{0.49}Ga_{0.51}P$ narrowband blu-ray detector

Thus, we can translate the photocurrent signal generated by the photoelectric detector into a digital voltage. It is desirable for its ease of use and ability to package both the photoelectric detector chip and front transresistance amplifier chip in the same package. In free-space optical communication applications, in order to ensure enough sensitivity and adaptability in the system, a larger reception area is often required. However, a large reception area means a bigger capacity, so the adverse influence from the RC time constant to the system response speed should be considered. An effective method is to maintain a small-size photoelectric detector and to use an optical element, such as a condenser lens or large Fresnel lens, to ensure good reception. For optical fiber communication applications, a photoelectric detector's photosensitive area is generally greater than the fiber core diameter (about 10 mm); thus, only a small area of the photoelectric detector can be used. However, a light-sensitive surface diameter of 50 mm is usually enough; thus, the capacitance is very small at about 1 pF level. On the other hand, the capacitance is still very large in free-space optical communication, so we will need a photodetector area large enough to make sure the system has a certain tolerance, even with optical elements. For the combination of the photodetector and preamplifier, the response bandwidth is an inverse ratio to the square root of the input capacitance. We can do a simple estimate in this case—when the photoelectric detector capacitance increases by 100 times, the response bandwidth will only be one-tenth of the original nominal value. Thus, we must consider the capacitance effect in a practical circuit. At the same time, circuit stability problems caused by the input capacitance also needed to be considered. In addition, the wavelength of the visible light band is shorter, compared with the 1.3-mm and the 1.55-mm near-infrared wave band optical fiber communication application, and the corresponding photon energy is higher. This means that the photoelectric detector degree response will be lower under the same quantum efficiency. Take the blue band for example: the responsivity is just below one-third, compared with the near-infrared wave band. In order to achieve the same received power sensitivity, improvement of the preamplifier gain is also required, and this should be estimated at the receiver.

In the following section, we discuss the advantages of using a narrowband spectral response photoelectric detector in the visible light free-space optical communication system, along with some general device design considerations. In addition, we use the $In_{0.48}Al_{0.52}P/In_{0.49}Ga_{0.51}P$ narrowband blue light detector as an example to explore the issues on material and device design aspects. We do this by using the band-gap of the semiconductor materials and absorption characteristics to reach a narrowband detection effect. In fact, there are other solutions in order for the photoelectric detector to achieve a narrowband detection effect, such as the optical Enhanced Bragg resonant cavity detector, band resonance effect of a narrowband detector, the introduction of a narrowband filter structure during the manufacturing of the device. For visible light communication, the blue band is not the only channel. When we use RGB-LED, red light components in the white light may be the better optical channel. This is because the cutoff wavelength of the $In_{0.49}Ga_{0.51}P$ material corresponds to the red light band (~ 650 nm), and there are also reports of similar devices in this aspect. Therefore, we can get a suitable red light narrowband detector by appropriately tailoring the response optical spectral and by designing the material combination of both $In_{0.49}Ga_{0.51}P$ and GaAs.

4.3   Blu-Ray Filters

4.3.1   An Overview

There are still a large number of disturbance light signals, which are much larger than the blue light in the visible light communication environment. In addition to the blue band communication signal with a peak at 456 nm, which includes a blue signal excitation phosphor in the LED, the result is white light illumination above 500 nm, as shown in Fig. 4.13. This part of the light can also enter the PIN diode detector, forming the light response signal, which makes it impossible to distinguish. Usually, the PIN diode detector's light response is significantly stronger when the wavelength is over 500 nm, compared to the blue light, which is below 500 nm. Therefore, the noise that is received by the visible light reception system, without any processing, is much greater than the blue communication signal itself. In addition to a strong interference from other visible light sources and sunlight, the system cannot achieve blue signal detection and communication. Not only does the strong visible noise need to be treated, but also the stray light in the near-infrared region needs to be filtered completely from the light-receiving system. Otherwise, it will form a strong noise when entering the light-receiving system, because the response peaks of commonly used PIN diodes are often in 800–900 nm near infrared, which is outside the visible light band.

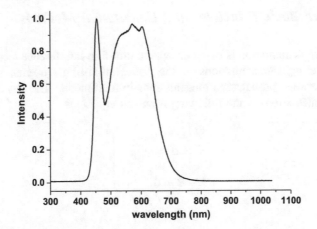

Fig. 4.13 White LED light spectrum in the visible light communication system

In order to avoid the blue signal being drowned out by loud noise and failing, it is necessary to add a blue light film for filtering before all of the light rays enter the detector. Its purpose is to filter all the stray light from the other bands and let the blue light for the optical communication transfer cleanly through. For the above-described white light LED, blue filters with the following characteristics need to be designed for filtering.

For this kind of LED, the blue band near 450 nm needs to be filtered out, while also closing off the other bands. The specific requirements for the design of its blue membrane are as follows:

(1) For the blue band of 420–478 nm, its average transmittance should not be less than 90%, to ensure that the blue light signal near 456 nm is transmitted completely into the detector.
(2) For the other visible light and near-infrared bands of 500–1000 nm, their average transmittance should not be more than 1%, to ensure that other visible light, including sunlight and all of the stray light near infrared, is filtered out.

That is to say, the blue band's transmittance should be as high as possible, while the transmittance of the other visible light and near-infrared light should be as low as possible, thus achieving a band-pass filter in the blue band. We expect the filter to create a useful blue light signal from the lot of noise in order to reach the available SNR. Ultimately, the integration of visible lighting and communication is achieved.

4.3.2 The Basic Principles and Calculation Methods

Blue filter implementation is based on optical thin-film interference technology. By studying the light transmissions in the optical thin-film systems, we are also studying the wave propagation transmissions in multimedia. The physical basis of its design calculations is the following Maxwell equations:

$$\left.\begin{aligned} \nabla \cdot D &= \rho \\ \nabla \times E &= -\frac{\partial B}{\partial t} \\ \nabla \times H &= j + j_D \\ \nabla \cdot B &= 0 \end{aligned}\right\}$$

(4.14)

$$\left.\begin{aligned} D &= \varepsilon E \\ B &= \mu H \\ j &= \sigma E \end{aligned}\right\}$$

(4.15)

We commonly use the transfer matrix method and optical admittance to calculate the reflection and transmission characteristics of the optical thin films in the actual calculation. The transfer matrix method is to expand the grid position of the electromagnetic fields in real space and transform the Maxwell equations into a transfer matrix form. It then becomes the problem of solving the eigenvalues. The essence is to convert the Maxwell equations into the transfer matrix, which is the transfer matrix method's modeling process. The details of the process are as follows:

To begin, use the Maxwell equations to solve the electric and magnetic fields between the two adjacent levels, so you can get the transfer matrix. Then, promote the single-layer conclusion to the entire media space. Thus, you can calculate the transmission and reflection coefficients of the entire multimedia.

Some characteristics of the transfer matrix method are fewer matrix elements, a small computation amount, and a fast speed. Generally, the key is to solve the matrix element. This method is especially suitable for the dielectric film of the periodically alternating multilayer.

First, we study the film's transmission and reflection characteristics in a single-layer film. For example, the structure diagram is shown in Fig. 4.14.

Using the electric field and magnetic field boundary conditions, we can see:

$$\begin{bmatrix} E_0 \\ H_0 \end{bmatrix} = \begin{bmatrix} \cos \delta_1 & \frac{i}{\eta_1} \sin \delta_1 \\ i\eta_1 \sin \delta_1 & \cos \delta_1 \end{bmatrix} \begin{bmatrix} E_2 \\ H_2 \end{bmatrix}$$

(4.16)

Among which, δ_1 is the thickness of the phase, $\delta_1 = \frac{2\pi}{\lambda} n_1 d_1 \cos \theta_1$, and θ_1 is the incident angle.

For the p component, $\eta_1 = n_1 / \cos \theta_1$ and for the s component, $\eta_1 = n_1 / \cos \theta_1$.

Fig. 4.14 Structure diagram
of the single-layer films

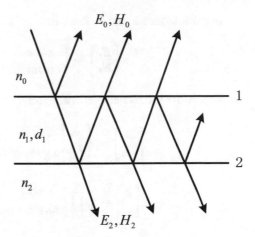

Assume that $\begin{bmatrix} B \\ C \end{bmatrix} = \begin{bmatrix} \cos\delta_1 & \frac{i}{\eta_1}\sin\delta_1 \\ i\eta_1\sin\delta_1 & \cos\delta_1 \end{bmatrix}\begin{bmatrix} 1 \\ \eta_2 \end{bmatrix}$, and this is the characteristic equation of the thin film.

So $M_1 = \begin{bmatrix} \cos\delta_1 & \frac{i}{\eta_1}\sin\delta_1 \\ i\eta_1\sin\delta_1 & \cos\delta_1 \end{bmatrix}$ is referred to the film characteristic matrix, and it contains all of the useful parameters of the thin film.

The admittance of the membrane layer and its basal combination is: $Y = C/B$.

So, the energy reflectance is: $R = \left(\frac{\eta_0 - Y}{\eta_0 + Y}\right)\left(\frac{\eta_0 - Y}{\eta_0 + Y}\right)^*$.

And its transmittance is: $T = \frac{4\eta_0\eta_2}{(\eta_0 B + C)(\eta_0 B + C)^*}$.

Similarly, for the multilayer thin-film structure shown in Fig. 4.15, we can also use the transfer matrix method to calculate the transmission and reflection characteristics.

In the interface 1 and 2, there is:

$$\begin{bmatrix} E_0 \\ H_0 \end{bmatrix} = \begin{bmatrix} \cos\delta_1 & \frac{i}{\eta_1}\sin\delta_1 \\ i\eta_1\sin\delta_1 & \cos\delta_1 \end{bmatrix}\begin{bmatrix} E_2 \\ H_2 \end{bmatrix} \tag{4.17}$$

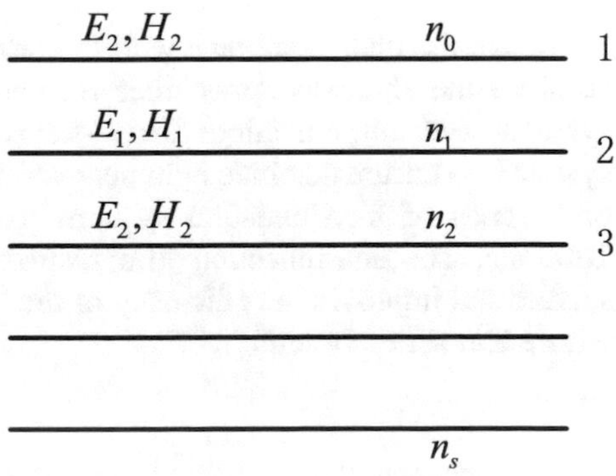

Fig. 4.15 Diagram of the
multilayer thin-film structure

E_2, H_2	n_0	1
E_1, H_1	n_1	2
E_2, H_2	n_2	3
	n_s	

In the interface 2 and 3, there is:

$$\begin{bmatrix} E_1 \\ H_1 \end{bmatrix} = \begin{bmatrix} \cos \delta_2 & \frac{i}{\eta_1} \sin \delta_2 \\ i\eta_2 \sin \delta_2 & \cos \delta_2 \end{bmatrix} \begin{bmatrix} E_3 \\ H_3 \end{bmatrix} \tag{4.18}$$

Repeat the process above, and finally:

$$\begin{bmatrix} E_0 \\ H_0 \end{bmatrix} = \left\{ \prod_{j=1}^{K} \begin{bmatrix} \cos \delta_j & \frac{i}{\eta_j} \sin \delta_j \\ i\eta_j \sin \delta_j & \cos \delta_j \end{bmatrix} \right\} \begin{bmatrix} E_{K+1} \\ H_{K+1} \end{bmatrix} \tag{4.19}$$

$$\begin{bmatrix} B \\ C \end{bmatrix} = \left\{ \prod_{j=1}^{K} \begin{bmatrix} \cos \delta_j & \frac{i}{\eta_j} \sin \delta_j \\ i\eta_j \sin \delta_j & \cos \delta_j \end{bmatrix} \right\} \begin{bmatrix} 1 \\ \eta_{K+1} \end{bmatrix} \tag{4.20}$$

The admittance of the multilayer film system is: $Y = C/B$.

Matrix $M_j = \begin{bmatrix} \cos \delta_j & \frac{i}{\eta_j} \sin \delta_j \\ i\eta_j \sin \delta_j & \cos \delta_j \end{bmatrix}$ is called the characteristic matrix of the j film, where δ_j is the thickness of the phase, and $\delta_j = \frac{2\pi}{\lambda} N_j d_j \cos \theta_j$

$$\eta_j = \begin{cases} n_j / \cos \theta_j, & p \text{ Polarization component} \\ n_j \cos \theta_j, & s \text{ Polarization component} \end{cases} \tag{4.21}$$

The reflectance of the multilayer film expression is the same:

$$R = \left(\frac{\eta_0 - Y}{\eta_0 + Y} \right) \left(\frac{\eta_0 - Y}{\eta_0 + Y} \right)^* \tag{4.22}$$

And the transmittance is: $T = \frac{4\eta_0 \eta_{K+1}}{(\eta_0 B + C)(\eta_0 B + C)^*}$

4.3.3 Blu-Ray Filter Design

To design the blue membrane with the proposed targets in Sect. 4.3.1, we need to combine the shortwave-pass filter membrane system, a series of high anti-film systems, and anti-reflection film systems. The shortwave-pass filter membrane system is to ensure that blue light near 456 nm can be transmitted. The combination of a series of high anti-film systems expands the high anti-area from 500 to 1000 nm. The anti-reflection film reduces the reflection losses of the substrate surface and improves the efficiency of the blue signal reception, on the basis of the whole film series system.

Fig. 4.16 Waveforms and parameters of the band-pass filter

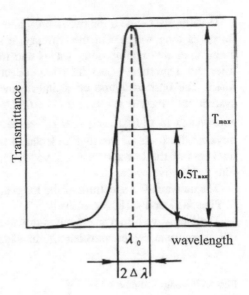

(1) Band-pass filtering

For the band-pass filter, a band pass usually requires a flat top and a high transmittance. The highly reflective and low-transmission cutoff requires the transmittance to be as low as possible, and the range to be as wide as possible, on both sides of the band pass. The waveforms and schematic parameters of the band-pass filter are shown in Fig. 4.16. The main parameters characterizing the band-pass filter are:

λ_0 The center wavelength or peak wavelength.

T_{max} The transmittance of the center wavelength, namely peak transmittance.

$2\Delta\lambda$ The wavelength width when the transmittance is half of the peak transmittance. We can use $2\Delta\lambda/\lambda_0$ to relatively represent the full width at half of the maximum.

There are two common types of band-pass filter structures:The first one is the simplest form of a FP interferometer, which consists of the cavity layer with high-reflection layers on both sides. This kind of structure, with its spectral characteristics, can achieve a very narrow passband; however, the cutoff frequency band is narrow as well. Therefore, we often need a filter to broaden the cutoff frequency band and increase the cutoff depth. The other band-pass filter combines a longwave film and a shortwave film. The spectral characteristics of this structure are that it can get a wider cutoff frequency and a deeper cutoff degree, but because it is not easy to get a narrow passband, it is often used to obtain a broadband filter.

For the visible light communication blu-ray film involved in this paper, the band pass (420–478 nm) is wide ($\sim$60 nm), but the cutoff bandwidth is quite wide (500–1000 nm). Therefore, we need a band-pass overlap system constructed by a longwave film system and shortwave film system.

Usually, we define a longwave-pass filter as a filter that inhibits shortwaves and transmits long waves. On the contrary, a shortwave-pass filter is defined as the cutoff filter that inhibits long waves and transmits shortwaves. A longwave-pass filter and a shortwave-pass filter are collectively referred to as interference-cutoff filters. The basic structure of an interference-cutoff filter is a $\lambda/4$ multilayer film system. Its structure is $\frac{H}{2} LHLH \cdots HLHL \frac{H}{2}$ or $\frac{L}{2} HLHL \cdots LHLH \frac{L}{2}$, which can be abbreviated to $\left(\frac{H}{2} L \frac{H}{2}\right)^s$ and $\left(\frac{L}{2} H \frac{L}{2}\right)^s$, respectively. Among them, H represents the physical thickness of the high-reflectance layer $\lambda/4 n_H$, L represents the physical thickness of the low reflectance layer $\lambda/4 n_L$, and s represents the number of the film's repetitive cycle.

The transmission spectrum of the long- and shortwave-pass filter wave is shown in Figs. 4.17 and 4.18, respectively.

The longwave film system and shortwave film system will overlap by selecting the appropriate center wavelength. In addition, the waveforms of the two film

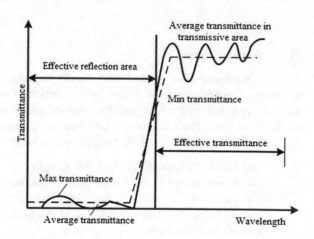

Fig. 4.17 Longwave-pass filter

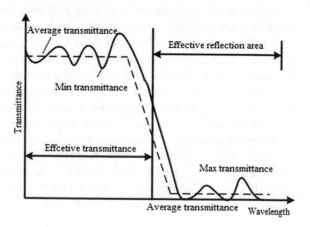

Fig. 4.18 Shortwave-pass filter

systems overlap. Their high transmittance areas coincide and constitute the band pass, and then, other parts coincide and constitute the cutoff band. Finally, they form the band-pass filter.

(2) The cutoff region extensions outside the band

Because of the bandwidth limitations formed by the simple superposition of the longwave-pass filter and shortwave-pass filter, the bandwidth cannot meet the design demand. In addition, the transmittance of the cutoff area is still high, so it is an incomplete cutoff. Therefore, we need to further optimize the design and extend the cutoff area. A specific method for this is to superimpose a different high-reflection film system, making it covered from 500 to 1000 nm, and with a cutoff depth below 1%.

Usually, the medium-to-high-reflective film system uses $\lambda/4$ film stacks $(HL)^n$, alternating between high- and low-refractive index materials. In addition, the thickness of each layer is $\lambda_0/4$, where n is logarithmic of the reflection plate (HL). When the film's reflection light beam, from of all of the interfaces, comes back to the surface, they will be in the same phase, resulting in a constructive interference. When the medium-to-high-reflection film system is more than ten layers, the reflectance can be up to 99%.

The reflection width of a single high-reflective film stack only depends on the difference between the refractive index of the high-refractive index film and the low-refractive index film. The greater the difference, the larger the width of the band will be. However, the actual materials that can be selected present a limited range, which significantly limits the difference between a high-refractive index film and a low-refractive index film. Particularly in the visible, near-infrared band, the difference is often less than 1, so that a single high-reflectance area of $\lambda/4$ film stack is very limited in width. In many applications, the high-reflectance region is not wide enough and cannot meet the requirements; therefore, we will need to expand the width of the high-reflectance region.

There are two common methods of broadening. One method is to make the film thickness of each layer progressively increase. This method's purpose is to ensure that any wavelength in a wide area within the multifilm layer has a sufficient number of film layers. Its optical thickness is very close to $\lambda/4$ and gives a high reflectance in a very wide band. The usual practice is to fix the refractive index of each layer, as well as optimize the thickness of each layer, automatically by a computer design. It is easy to obtain a broadband reflective film system with an uneven thickness; however, the production of an uneven thickness film system is quite difficult.

Another broadening method is by superimposing another high-reflective multi-layer film whose center wavelength is different on a $\lambda/4$ high-reflective multilayer film. With this method, you need to directly splice the high reflectivity tape of different bands to extend its high-reflectance region.

(3) The compression of the band-pass ripple

After solving the extension problem in the cutoff area, there is another problem that needs to be resolved. The problem is that the simple combination of different film systems will lead to a band-pass ripple, which means that the transmittance of a considerable area of the passband will not be high. Specifically, the transmittance of the blue light signal efficiency will be low and will influence the light-receiving efficiency. Therefore, we need to try to compress the ripple amplitude, to further improve the waveform of the band-pass filter, and to increase the average transmittance.

There are many different ways to compress a band-pass ripple. The simplest way is to select a symmetric combination, so that the refractive index of the substrate n_1 is close to the equivalent refractive index of the passband n_2. As long as the reflection loss on the surface of the substrate is not too high, it is a very good method to use. However, for different substrate materials, practical film material is not necessary to have a suitable refractive index equivalent.

Another method is to change the film thickness within the fundamental period, so that the equivalent refractive index is similar to the refractive index of the substrate. To make such method effective, it is required that the substrate maintains a low reflectance, namely that the substrate must have a low refractive index. In addition, with the visible light region, the glass substrate material must be satisfactory.

The common way to suppress a band ripple is to plate matching layers on both sides of the symmetrical film system, so the equivalent refractive index can be simultaneously matched to the substrate and the incident medium. We can insert a $\lambda/4$ layer, whose modified admittance is η_3 between the substrate and the symmetrical film system, and insert a $\lambda/4$ layer, whose modified admittance is η_1 between the incident medium and the symmetrical film system. The parameters simply need to satisfy: $\eta_3 = \sqrt{\eta_g E}$, and $\eta_1 = \sqrt{\eta_0 E}$, so the combination admittance of the symmetrical film system is: $Y = \eta_1^2 \eta_3^3 / E^2 \eta_g$. Thus, the reflectance is:

$$R = \left[\eta_0 - \eta_1^2\eta_3^2/(E^2\eta_g)\right]^2 \Big/ \left[\eta_0 + \eta_1^2\eta_3^2/(E^2\eta_g)\right]^2 \tag{4.23}$$

When $\eta_1^2\eta_3^3 = \eta_0/\eta_g$, we get $R = 0$.

(4) The aid of anti-reflection film

Finally, because of the superposition of a series of film systems as described above, to realize the design of the entire blue-layer film filter, we need almost one hundred films. In addition, we must take the actual materials absorption into account, which will result in a corresponding absorption loss that cannot be avoided. Furthermore, the surface of the blue filter film or the back substrate will have a certain degree of reflection when the signal light is incoming. This happens because the refractive index is larger than the air, which results in further signal loss. By introducing anti-reflection film, we can reduce the reflection and improve the final efficiency of

Fig. 4.19 A single-layer
anti-reflection film structure

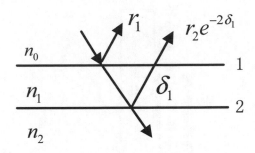

the transmittance and reception of the signal. Also, we can place corresponding anti-reflection film on the back of the blue filter to improve the transmittance of the blue signal path, namely the blue light signal's receiving efficiency. The blue light signal comes through the anti-reflection film surface, enters the light-receiving detector, and then finally becomes the communication signal we need. For a single layer of anti-reflection film, the structure is shown in Fig. 4.19.

We can calculate reflectance according to the transfer matrix method:

$$R = \left(\frac{n_0 - n_1^2/n_2}{n_0 + n_1^2/n_2} \right)^2 \tag{4.24}$$

When the film refractive index is $n_1 = \sqrt{n_0 n_2}$ and reflectance $R = 0$, this is the best anti-reflection effect.

However, in practice it is difficult for the refractive index of the film material to meet the above formula. We usually use a multilayer anti-reflection film in order to achieve a better anti-reflection effect. Usually, a multilayer anti-reflection film is composed of $\lambda/4$ layer or half-wavelength layer, which can be seen as the improvement of the double anti-reflection film $(\lambda/4 - \lambda/2)$ W-shaped membrane and $(\lambda/4 - \lambda/4)$ V-shaped membrane. We can use the transfer matrix method to optimize the anti-film multilayer system as well.

4.3.4   Design Examples

After the above steps of design and optimization, the final blue filter can be better than the design index. In the following examples, we take SiO_2 film material $(n = 1.46)$ and Ta_2O_5 film material $(n = 2.16)$, which are most commonly used in the visible light absorption band. We designed a blue-band-pass filter film used in the white LED communication system, and the results are shown in Fig. 4.20. From the figure, we can see that the blue-band-pass filter covers the whole blue part of the white LED, which has a signal peak of 456 nm. Any white light interference above 500 nm stimulated by phosphor is all filtered off by the 500–1000-nm film's filter

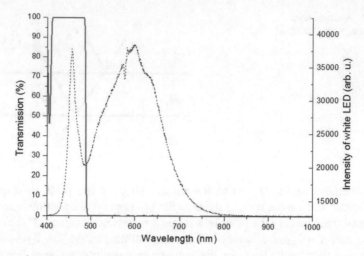

Fig. 4.20 White LED spectrum of visible light communication (dotted line) and the transmission spectrum of the designed blue filter (solid line)

region. Therefore, ultimately only the blue signal light reaches the light-receiving detector, which means we were able to filter out the blue light signal from a lot of noise light. As a result, we need to make sure of the signal-to-noise ratio (SNR) and the possible integration of visible lighting and communication.

4.3.5 Preparation

There are many methods for producing thin films. However, we generally produce them under certain vacuum conditions, due to a high demand for quality and the blue filter's optical film thickness control. We often use physical vapor deposition (PVD), chemical vapor deposition (CVD), thermal evaporation, sputtering, and ion plating. The thickness control methods of the film include: visual monitoring, the optical extreme value method, a quartz crystal monitor, single-wavelength monitoring, or wide spectrum scanning, so that we can ensure the exact film coating of each layer.

4.4 The Detector Circuit Design

After the PIN converts the received optical signal into an electrical signal, we require the reception electrical module, which is used for demodulation and the digital signal processing, to recover the original data. Generally, the demodulator

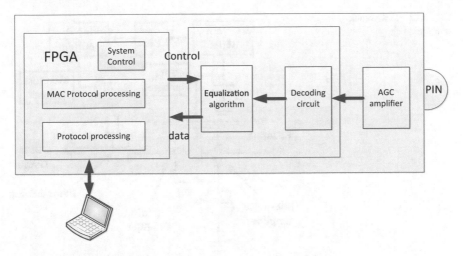

Fig. 4.21 Main structure of the receiving detector circuitry module

chip includes a low-noise amplifier, the automatic gain control circuit, a clock recovery circuit, a digital signal processing module, and other components.

The main structure of the receiving detector circuitry module is shown in Fig. 4.21. First, the received PIN signal will pass through a low-noise AGC amplifier, then transfer to the demodulator module for demodulation. After the demodulation of the OFDM signal, there will be signal processing and equalization on the recovered signal. Next, MAC parse is conducted by FPGA, and then finally, the original data is recovered. The demodulator chip receiver is designed to have a high sensitivity and response speed, as well as a low complexity, so that it can be applied to a high-speed indoor visible light communication system. The key technologies include the following.

4.4.1 Adaptive Receiver Technology

In actual use, changes in the distance of an LED communication system will make the received signal power change, which in turn will affect the decision level setting. This will inevitably lead to signal error. To solve this problem, we need to design an adaptive receiver module in the receiving circuit. Through adaptive signal power control of the receiver module, the received signal power that is inputted into the back-end signal processing module will be within an acceptable range. For a weak signal, we use filtering and amplification to realize a correct demodulation. The adaptive receiving module consists of an AGC circuit and a low-noise amplifier.

Among them, the automatic gain control (AGC) circuit is used for monitoring the demodulated signal, as well as controlling the low-noise amplifier's (LNA) gate

Fig. 4.22 AGC block diagram

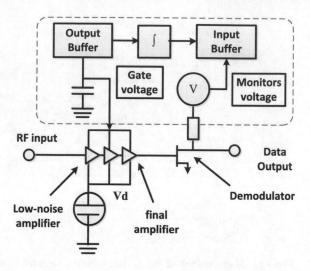

voltage, so that additional power cannot enter the final amplifier, and thus, an overload is avoided [1]. The AGC block diagram is shown in Fig. 4.22.

The AGC loop is a closed-loop negative feedback electronic circuit, which can be divided into two parts: a gain-controlled amplifier circuit and the voltage form control circuit. The gain-controlled amplifier circuit is located in the forward amplification paths, and the gain will change with the control voltage. The basic components of the control circuit are the AGC detector and the low-pass smoothing filter. However, sometimes it also includes a DC amplifier circuit and other components. The output signal is amplified by the detector and then goes through a filter to remove the low-frequency modulation components. This process generates the voltage to control the gain of the voltage-controlled amplifier. When the input signal increases, the gain of the amplifier circuit will decrease. In turn, this means that the output signal's amount of change is significantly smaller than the input signal's amount of change, which achieves the purpose of automatic gain control. After the gain adjustment, the receiving signal goes through a low-noise amplifier to achieve an adaptive receive process. Afterward, we can do further judgments and demodulation.

4.4.2   The Clock Extraction and Recovery Circuit

The receiver needs to decode the signals that are transmitted by the LED. In order to complete the decoding operation properly, we need to synchronize and extract the clock at the receiving end, and then we can find the accurate beginning of the data. There are a variety of clock synchronization receiver schemes, including a full network clock synchronization method, a remaining clock synchronization method, an adaptive clock synchronization method. In the designed white LED

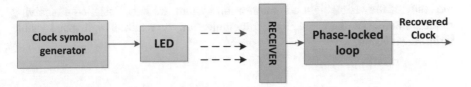

Fig. 4.23   Method of adaptive synchronization

communication system, there is no clock source signal from the transmission side to the receiving end, so we need to adopt a method of adaptive synchronization. Its process is shown in Fig. 4.23.

We need to send a clear time stamp from the transmission end to the receiving end. The time stamp is included in the data packet, and it can make both the receiving end's local clock and the transmitting end's clock synchronize. Since there is no common reference clock, the clock in the receiving end can only keep synchronization by restoring the clock in the transmitting end. This kind of synchronization method is similar to that the synchronization information is periodically inserted into the bit stream at the transmitting end, and then, the synchronization information is detected at the receiving end, so it then generates a reference signal to drive the phase-locked loop. The phase-locked loop can restore the transmitting end's clock. A PLL is used to estimate and compensate for the frequency drift between the sending clock oscillator and the receiving clock oscillator. By this method, the receiver clock can achieve synchronization and extraction.

4.4.3   Receiver Equalization Technology

In order to expand the bandwidth of the LED, we can add an equalizer into the receiver circuit. By doing this, the signals' frequency response attenuation can be compensated after equalization and the bandwidth can be expanded. To employ the equalization method described above, the modulation bandwidth of the blue channel system should be around 14 MHz, and the modulation bandwidth of the VLC system, after using equalization techniques, should be extended to 50 MHz. Under the condition of this bandwidth, we have achieved a transmission rate up to 100 Mbit/ s with NRZ-OOK, and BER $< 10^{-9}$.

4.5   Summary

This chapter analyzes the silicon PIN photodetector, the narrowband blue detector, the blue filter, and a detector circuit in the system's receiving module, from the perspective of visible light communication system receiver technology. Through the study of the visible light communication system, we have designed a receiving module suitable so that visible light communication can realize a high-speed visible light transmission.

References

1. Li, R., Wang, Y., Tang, C., et al.: Improving performance of 750-Mb/s visible light communication system using adaptive Nyquist windowing. Chin. Opt. Lett. **11**(8): 080605 (2013)
2. Zhang, Y.G., Gu, Y.: Al(Ga)InP-GaAs photodiodes tailored for specific wavelength range. In: Yun I (ed.) Photodiodes-from Fundamentals to Applications. Intech. ISBN: 978-953-51-0895-5 (2012)
3. Gu, Y., Zhang. Y.G., Li, H.: Gas source MBE growth and doping characteristics of AlInP on GaAs. Mater. Sci. Eng. B. **131**(1), 49–53 (2006)
4. Gu, Y., Zhang, Y.G., Li, A.Z., et al.: Optical properties of gas source MBE grown AlInP on GaAs. Mater. Sci. Eng. B **139**(2), 246–250 (2007)
5. Zhang, Y.G., Li, C., Gu, Y., et al.: GaInP-AlInP-GaAs blue photovoltaic detectors with narrow response wavelength width. IEEE Photon. Technol. Lett. **22**(12), 944–946 (2010)
6. Li, C., Zhang, Y.G., Gu, Y., et al.: Gas source MBE grown Al0.52In0.48P photovoltaic detector. Cryst. Growth **323**(1), 501–503 (2011)
7. Zhang, Y.G., Gu, Y., Zhu, C., et al.: AlInP-GaInP-GaAs UV-enhanced photovoltaic detectors grown by gas source MBE. IEEE Photon. Technol. Lett. **17**(6), 1265–1267 (2005)
8. Kuang, C., Zhang, Z.: Transfer matrix method analysis of light transmission properties of one-dimensional photonic crystals. Laser Mag. **4**(24), 38–39 (2003)
9. Tang, J., Gu, P., Liu, X., Li, H.: Modern optical thin film technology. Zhejiang University Press (2006)

Chapter 5
The Modulation Technologies of Visible Light Communication

In visible light communication systems, the LED's modulation bandwidth is quite limited. The 3-dB modulation bandwidth of the current commercially available LEDs is only few MHz. In order to improve the VLC system's transmission data rate, apart from the LED structure and the drive circuit design, selecting the appropriate modulation format is one of the most important ways to expand the modulation bandwidth. OOK modulation and PPM/PMW modulation have an advantage regarding the system complexity; however, DMT modulation, OFDM modulation, and CAP modulation have better spectral efficiency and can overcome multipath effects [1]. This chapter will introduce various modulation format principles and their implementations, advantages and disadvantages in visible light communication systems.

5.1 OOK Modulation Format

5.1.1 The Principle of the OOK Modulation Format

On-off keying (OOK) modulation format, also known as binary amplitude-shift keying (2ASK), uses a unipolar NRZ code sequence to control the opening and closing of a sinusoidal carrier. In OOK modulation, the amplitude of a carrier changes in only two states, corresponding to the binary information "0" or "1".

Amplitude-shift keying utilizes the amplitude change of a sinusoidal carrier to transmit information, while the frequency and initial phase of the sinusoidal carrier remain unchanged. For binary amplitude-shift keying, when transmitting the symbol "1", the amplitude of the sinusoidal carrier is taken as A_1, and when transmitting the symbol "0", the amplitude is A_2. The different symbols can be distinguished by the amplitude of the carrier, which can be expressed as (Sect. 5.1.1):

© Tsinghua University Press, Beijing and Springer-Verlag GmbH Germany 2018   91
N. Chi, *LED-Based Visible Light Communications*, Signals and Communication Technology, https://doi.org/10.1007/978-3-662-56660-2_5

$$e_{2ASK}(t) = \begin{cases} A_1 \cos(\omega_c t + \varphi), & \text{send "1" at probability } P \\ A_2 \cos(\omega_c t + \varphi), & \text{send "0" at probability } 1 - P \end{cases} \quad (5.1)$$

Assuming $A_1 = A$, $A_2 = 0$ and the initial phase $\varphi = 0$, (5.1) can be expressed as:

$$e_{2ASK}(t) = \begin{cases} A_1 \cos(\omega_c t + \varphi), & \text{send "1" at probability } P \\ 0, & \text{send "0" at probability } 1 - P \end{cases} \quad (5.2)$$

The waveform of OOK modulation format is shown in Fig. 5.1. The carrier is changing under the control of a binary baseband signal $s(t)$. In OOK, "1" or "0" is represented by the absence of voltage.

According to the basic principles of the binary amplitude-shift keying, a general expression of 2ASK signals can be written as:

$$e_{2ASK}(t) = s(t) \cos \omega_c t \quad (5.3)$$

Here, $s(t) = \sum_n a_n g(t - nT_s)$ is a binary unipolar baseband signal. For simplicity, we typically assume that $g(t)$ is a rectangular pulse at the height of 1 and the width of Ts, and a_n is a binary symbol sequence, whose value is 1 or 0.

According to the basic principles of OOK, we can get two modulation methods: "analog modulation" and "keying modulation." The schematic diagram of the modulators is shown in Fig. 5.2.

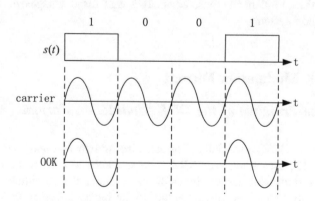

Fig. 5.1 Waveform of the OOK signals

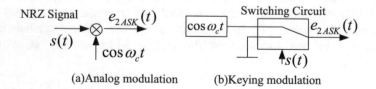

(a)Analog modulation　　　(b)Keying modulation

Fig. 5.2 Schematic diagram of the modulators

Since the anti-noise performance of OOK is not as good as other modulation methods, this modulation format has not been used with current satellite communications or digital microwave communications. However, due to its simple implementation, OOK is usually utilized in visible light communication systems to test the system's performance.

5.1.2 The BER Performance of OOK

A transmitter transmits a pulse to indicate "1" at the duration time of $1/R_b$ (R_b = bit rate) and a density of $2P$ (P is the transmission rate) and then indicates "0" when there is no pulse. The modulation bandwidth of the OOK signal is $R_b = 1/T$, which is the reciprocal of the pulse width.

The transmitter transmits one of the negative signals of L $\{x_1(t), x_2(t), \ldots, x_L(t)\}$ at the data rate of R_b every $T = \log_2(L/R_b)$, and the channel additive white Gaussian noise power spectrum is N_0. In order to prevent inter-symbol interference, each signal is limited to the interval $[0, T)$. The signal set is satisfied with $x(t) \geq 0$ and $\overline{x(t)} \leq P$ (where P is the optical transmitter power limit); thus, the average signal power is $\frac{1}{L}\sum_i \overline{x(t)} \leq P$. For simplicity, we use the highest SNR and assume that the BER is controlled by the two adjacent signals, namely

$$\text{BER} = Q\left(\frac{d_{\min}}{2\sqrt{\sqrt{N_0}}}\right) \tag{5.4}$$

where $d_{\min}$ is the minimum Euclidean distance between the modulated signals:

$$d_{\min}^2 = \min_{i \neq j} \int \left[X_t(t) - X_j(t)\right]^2 dt \tag{5.5}$$

The Q function is correlated with the error function (erf), which is defined as:

$$Q(x) = \frac{1}{\sqrt{2x}}\left[\frac{1}{2} - \frac{1}{\sqrt{x}}\int_0^x e^{-t^2} dt\right] \tag{5.6}$$

The BER performance of OOK signal is:

$$\text{BER}_{\text{OOK}} = Q\left(\frac{P}{\sqrt{N_0 R_b}}\right) \tag{5.7}$$

The power to achieve the desired BER is:

$$P_{\text{OOK}} = \sqrt{N_0 R_b}Q^{-1}(BER) \tag{5.8}$$

And the required power of the other modulation schemes to achieve the same BER is:

$$P = \left(\frac{d_{\text{OOK}}}{d_{\min}}\right)P_{\text{OOK}} \tag{5.9}$$

5.1.3 System Implementation and Waveform Testing

In the lighting, there are two types of white LEDs:

(1) Use separate red, green, and blue emission diodes (RGB-LED) and
(2) Use the blue LED with yellow phosphor.

The experiment setup is shown in Fig. 5.3. In the experiment, a commercially available blue LED with yellow phosphor is used as the optical source. Since the LED lamp has very wide radiation angle, focusing lenses are used in the free-space channel to focus sufficient light to be received for a photodiode (PD). Pseudorandom NRZ signals at different data rates are generated by a signal generator and then amplified through the amplifier. The AC amplifier utilized in the experiment is the conventionally designed two-stage class AB amplifier. The output of the amplifier is combined with a DC bias via a T-type bias tee to drive the LED module. The LED emits white light through the free-space propagation channel to a PD, which has a blue filter and a condenser at the front end. Next, the optical signal is detected by the PD and then sent to the transimpedance amplifier. At last, the signal is sent to the oscilloscope and the BERT to record the signal eye diagram and calculate the BER.

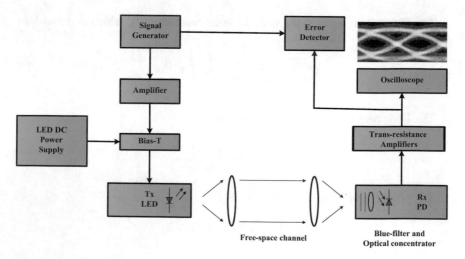

Fig. 5.3 Experimental setup of the OOK transmission system

(a)   (b)

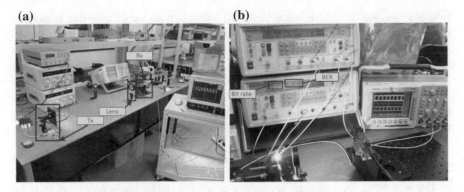

Fig. 5.4 **a** VLC experimental setup and **b** testing result of the 20-Mb/s VLC experimental system

The visible light communication system is built according to the schematic diagram, as shown in Fig. 5.4a. The system's transmission data rate is not very high, but it has laid a solid foundation for the following research work. The VLC system utilizes the OOK modulation, and a 20 Mb/s error-free transmission can be achieved without the focusing lens at the receiver side, as shown in Fig. 5.4b.

Among all of the intensity modulation/direct detection-based modulation technologies, OOK is considered the most commonly used technology in the digital optical transmission systems due to its simple implementation. In 2009, South Korea's Samsung and Oxford University utilized the OOK-NRZ modulation to achieve a 100 Mbit/s data transmission at a low BER. German Fraunhofer Communication Institute and Siemens have achieved 125 Mbit/s data rates using the OOK modulation and a PIN diode receiver, and 230 Mbit/s with an avalanche photodiode (APD).

5.2   The PPM and PMW Modulation Technologies

Pulse position modulation (PPM) was first proposed by Pierce JR and then applied to space communications. There are three basic methods to convert a digital sequence into a pulse sequence: changing the amplitude, changing the position, and changing the width of the pulse. The corresponding modulation methods are known as PAM modulation, PPM modulation, and PWM modulation. PPM modulation is to generate a PPM pulse signal by encoding and then modulates the signals in visible light communication. There is a certain relationship between the pulse position and the signal sample values in one period. In the PPM modulation-based system, the transmitted data determines the position of the pulse, while the PWM modulation is used to control the brightness of the LED in visible light communication. In many current literatures, researchers have combined both PPM and PWM.

Single-pulse position modulation (L-PPM) refers to a period of time that is divided into several time slots, and then, a single-pulse signal is transmitted at one time slot for data transmission. Single-pulse position modulation can be described as mapping an n-bit binary data to a single-pulse signal at a certain time slot period of the $N = 2^n$ time slots. The symbol interval is divided into N time slots, and the width of each time slot is T/N. If the n-bit data consists of $M = (m_1, m_2, \ldots, m_n)$, and the time slot position is referred as K, the single-pulse position modulation mapping relationship is (5.10):

$$K = m_1 + 2^* m_2 + \cdots + 2^{n*} m_n, \ m_n \in \{0, 1, \ldots, n - 1\} \qquad (5.10)$$

For a 4-PPM modulation, according to (5.10), (0,0) corresponds to the slot position "0", (1,0) corresponds to the slot position "1", (0,1) corresponds to the slot position "2", and (1,1) corresponds to the slot position "3". As shown in Fig. 5.5, the mapping is to create a one-to-one corresponding relationship, which is why modulation is unique.

Jang et al. [2] have proposed a PPM modulation scheme based on PWM. In this scheme, the PPM signal is added to the PWM dimming control signals when the data is transmitted and brightness control is operated. In order to express the brightness control in the PPM data transmission, the PWM duty ratio changes from 40 to 80%. The PWM frame rate is set to 1 kHz to reduce the flickers seen by the human eyes. The PPM data rate is set to 20 kbit/s, and the BER of 10^{-5} is achieved in the simulation. Compared with NRZ, PPM has a better power efficiency because DC and low-frequency spectral components are avoided. Moreover, PPM can maintain a dimming distribution in a constant average power. Therefore, for VLC systems, it is recommended to use a PPM based on the PWM dimming control.

Sugiyam et al. [3] combined L-PPM and PWM and have proposed a subcarrier pulse position modulation (SC-PPM) for visible light communication. Subcarrier modulation is reliable, because it is not affected by background light, such as fluorescent lamps. In SC-PPM, symbol intervals are divided into L time slots, and the optical signal is the subcarrier at the lth time slot. When subcarriers exist, the information is transmitted by the position. The average power of the transmitted signal is always the same and is independent of the data sequence. SC-4PPM is a standard modulation format for visible light modulation, its bit rate is 4.8 kbit/s, and the subcarrier frequency is 28.8 kHz. By choosing a sufficient high frequency of PWM, PWM and SC-PPM can be used simultaneously. SC-PPM is used to transmit

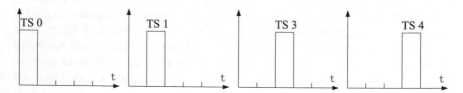

Fig. 5.5 Schematic diagram of 4-PPM

the data sequence, while PWM is used to control the brightness of the LEDs. A transmission waveform is generated via the SC-PPM waveform, multiplied by the PWM waveform. This approach has the following advantages: The PWM can be easily implemented by a digital processor without the need to use a complex analog circuitry, SC-PPM can be used for communication, and PWM is used for dimming and can be controlled independently. However, the frequency of PWM is too low and close to the subcarrier frequency; thus, the subcarrier will be cut off by the PWM waveform.

Although PPM is widely used and shows both good power efficiency and error performance, the PPM's bandwidth efficiency is very low and the pulse duration is very short. Moreover, at the receiver, symbols and time slot synchronization are also needed, which is not conducive to the realization of the system. Researchers have made an improvement based on L-PPM to obtain Differential Pulse Position Modulation (DPPM). The principle of DPPM is to start a new PPM symbol immediately after a time slot containing pulse and then remove the pulse-free time slot after the pulse time slot, thereby significantly increasing the throughput. DPPM always has higher spectral efficiency than PPM, and it does not require symbol synchronization. However, the DPPM's pulse duration width is very short; therefore, this system is also difficult to achieve. Huang et al. [4] have combined DPPM and PWM and proposed a differential pulse width modulation (DPPM + PWM) scheme. Compared to PPM, this scheme does not require symbol synchronization at the receiver and can effectively solve the narrow pulse problem in both DPPM and PPM. The simulation results show that DPPM + PWM has low power efficiency, but it is still higher than just PWM. Nevertheless, the error performance of DPPM + PWM is below PPM and DPPM, but clearly better than OOK. Compared with OOK, PPM, and DPPM, DPPM + PWM has higher bandwidth efficiency. Moreover, compared with PPM, DPPM + PWM does not require symbol synchronization. Therefore, DPPM + PWM can be substituted for PPM in many applications.

Multiple pulse position modulation (MPPM) is also developed on the basis of the PPM. MPPM's purpose is to map the n-bit binary data sequence into multiple pulses consisting of 2^n time slots, which can transmit multiple pulses within one frame period, while L-PPM can only send one pulse. MPPM is used to improve the bandwidth efficiency of PPM. Under the same condition, this approach would reduce the bandwidth to about half of a conventional PPM [5]. Siddique et al. [6] have noted that the information capacity of MPPM is affected by the total number of time slots and the number of the pulse time slots per symbol period. Therefore, the implementation of MPPM is much more complicated and its decoding is very difficult; thus, it is very suitable for tight bandwidth conditions. Kim et al. [7] have proposed a dimming system based on MPPM, which is realized by varying the number of optical pulses within the duration of one symbol. This system can achieve high spectral efficiency with relatively less optical power when the code length is increased.

Due to flicker effect during communication, Bai et al. [8] have proposed the overlapping pulse position modulation (OPPM). OPPM is defined as a special form of MPPM. OPPM can minimize the flicker effect during communication and achieve dimming control through the signal amplitude. OPPM has a large dimming range and can further improve the bandwidth efficiency of MPPM. It can also achieve high-rate communication, with a low flicker effect, at the cost of a variable current LED driver and accurate receiver synchronization. Wu et al. [9] have applied OPPM and PWM to indoor visible light communication and achieved a 400 Mbit/s transmission data rate by theoretical analysis, and a 50 Mbit/s bit rate is obtained via the experiment.

Although PPM is widely used in optical wireless communication systems, it is not suitable for VLC because it is difficult to control the LED's brightness. Kim et al. have proposed variable PPM (VPPM), which is a combination of PWM brightness control and a 2-PPM data transmission. Since VPPM is the conversion of 2-PPM, it carries only one bit of information in one symbol duration. However, simulations show that the performance of MPPM is better than VPPM.

Because PWM is used to traditionally control the brightness, the use of VPPM to transmit data will make the system design very complex. Therefore, Siddique et al. [6] have proposed a variable rate multipulse modulation (VR-MPPM), based on a white LED visible light communication system, to achieve the combined brightness control and data transmission. For VR-MPPM, the brightness control scheme depends on the number of time slots per symbol and the data rate depends on the number of pulses per symbol time slot. However, the defect of VR-MPPM is that the actual data transmission rate is not consistent at all brightness levels.

5.3 DMT Modulation Technology

Discrete multitone (DMT) modulation is one of the OFDM modulations, which mainly uses the inverse fast Fourier transform (IFFT) to convert a complex signal into the real signal. By this way, there is no need to use the IQ modulation, thereby reducing the complexity and cost of the system and thus making it suitable for a system with a low complexity, such as visible light communication systems.

5.3.1 The Principle of DMT Modulation and Demodulation

The schematic diagrams of a DMT modulation and demodulation are shown in Figs. 5.6 and 5.7.

The input data is divided into the N parallel subcarrier streams, and each subcarrier is modulated by the encoded high-order QAM complex symbols. Then, just like the OFDM modulation, these symbols are sent for DMT modulation, which is achieved by IFFT processing. To make all the DMT modulated data real, DMT modulation requires a $2N$-point IFFT transformation, and the values inputted into the IFFT transformation are needed to satisfy the Hermitian symmetry properties:

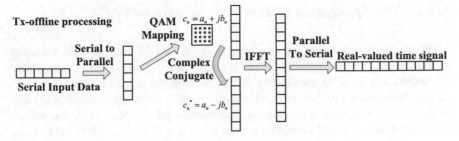

Fig. 5.6 Schematic diagram of DMT modulation

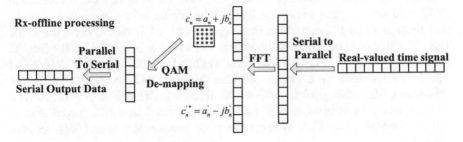

Fig. 5.7 Schematic diagram of DMT demodulation

$$C_{2N-n} = C_n^* \tag{5.11}$$

Here, $n = 1, 2, \ldots N - 1$, $\text{Im}\{C_0\} = \text{Im}\{C_N\} = 0$. That is, in order to ensure that the time domain signal is a real number through the IFFT transformation, a $2N$-point IFFT transformation is needed. Moreover, it is also required that the former half of the data for the $2N$-point IFFT conversion is a complex conjugate of the latter half. Therefore, the output after the $2N$-point IFFT is:

$$s_k = \frac{1}{\sqrt{2N}} \sum_{n=0}^{2N-1} C_n \exp\left(j2\pi k \frac{n}{2N}\right), k = 0, 1 \ldots, 2N - 1 \tag{5.12}$$

Here, S_k is the real-value signal consisting of $2N$ points.

At the receiver side, the DMT demodulation is achieved by a $2N$-point FFT transformation, which is inverse to the IFFT transformation. After the DMT demodulation, the symbols can be expressed as:

$$C_n = \frac{1}{\sqrt{2N}} \sum_{n=0}^{2N-1} s_k \exp\left(-j2\pi k \frac{n}{2N}\right), n = 0, 1 \ldots, 2N - 1 \tag{5.13}$$

It can be seen that, after FFT demodulation, the symbol C_n can be recovered.

5.3.2   The Application of DMT Modulation in VLC

DMT modulation not only has the advantages of OFDM such as high-frequency efficiency and anti-multipath effects, but also has a lower system complexity. Therefore, it is an ideal modulation format for high-speed visible light communication. References [10–12] have investigated DMT modulation-based VLC systems, revealing an increasing system transmission rate. In Ref. [12], the authors have made use of DMT modulation to achieve a 1.25 Gbit/s WDM VLC transmission using RGB-LED and APD, which is the highest data rate using DMT in VLC systems. The experimental setup is shown in Fig. 5.8:

In the experiment, a RGB-LED is utilized as the WDM optical source. Next, the DMT modulated signals generated by an arbitrary waveform generator (AWG), pass through an amplifier, and are then applied with a DC bias to drive the LED. The DMT signals consist of 128 subcarriers and have a bandwidth of 100 MHz. At the receiving end, an APD (the 3 dB bandwidth is 80 MHz) is used. The received signals are recorded by a storage oscilloscope for DMT demodulation. In the experiment, the BER performance of all three wavelengths is measured. The analysis and experimental results have demonstrated that DMT modulation is suitable for high-speed VLC systems due to its low complexity and high spectral efficiency.

5.4   OFDM Modulation Technology

Orthogonal frequency division multiplexing (OFDM) is a new and efficient modulation technique and is one of the multicarrier modulations. It can effectively resist multipath interference so that the signal can be reliably received, and it can also significantly improve the spectral efficiency [13]. In 1971, Weinstein and Eben proposed to realize the modulation and demodulation of OFDM by using the

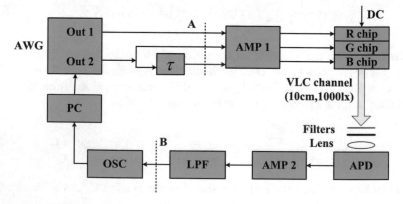

Fig. 5.8 Experimental setup of the DMT-based WDM VLC system

discrete Fourier transformation, which can simplify the oscillator arrays and the strict requirement of synchronization between the local carriers at the coherent receiver. Their work has laid theoretical preparation for the realization of all-digital OFDM modulation schemes. After the 1980s, with the development of digital signal processing (DSP) technology and the growth of high-speed data communication demands, the OFDM technique once again became a hot topic. OFDM technology has gained more and more attention because it has many unique advantages such as:

(1) High spectral efficiency. The spectral efficiency of OFDM is nearly twice as that of the serial communication system. It is important among wireless communications with limited spectrum resources.
(2) The ability to resist multipath interference and frequency selective fading. Because the data in OFDM system is spread between many subcarriers, which significantly reduces the symbol rate of each subcarrier, therefore the effect of multipath interference is weakened. In addition, the inter-symbol interference can be completely eliminated, even if a cyclic prefix is used as the guard interval.
(3) Dynamic subcarrier allocation technology to achieve the maximum bit rate. The transmission bit rate is maximized by selecting the various subcarriers, the number of bits in each symbol, as well as the power allocated to each subcarrier.
(4) A strong anti-fading capability by joint coding of each subcarrier. OFDM modulation technology itself has made use of channel frequency diversity. Additionally, the system performance can be further improved by the joint coding of each subcarrier.
(5) OFDM utilizes both IFFT and FFT to realize modulation and demodulation, which can be easily implemented by DSP.

The principle of OFDM is: The channel is first divided into a number of orthogonal subchannels. Then, high-speed serial signal streams are converted into parallel low-speed subdata streams. Because the bandwidth of each subchannel is only a small part of the original channel bandwidth, the channel equalization is relatively easy to be realized.

The basic idea of OFDM technology is: The high-speed serial data is converted into relatively low-speed parallel data to be modulated to each subcarrier. This parallel transmission scheme extends the pulse width of the symbol greatly and improves the resistance to multipath fading. Quadrature signals can be separated by using correlation techniques at the receiver, thus reducing the inter-subcarrier interference. The bandwidth of each subchannel is less than the channel's coherent bandwidth; therefore, it can be regarded as flat fading in the subchannel, thereby eliminating the inter-symbol interference. In traditional frequency division multiplexing, the spectrums of each subcarrier are nonoverlapping. In addition, large amounts of transmitting and receiving filters are required, thus significantly increasing system's the complexity and cost. Meanwhile, in order to reduce the cross talk between different subcarriers, a sufficient guard interval must be maintained, which will reduce the frequency efficiency of the system. However, the modern OFDM system uses digital signal processing techniques to generate and receive signals, which greatly simplifies the system's structure. Moreover, in order to improve spectral efficiency, each subcarrier is overlapped as shown in Fig. 5.9. These subcarriers should maintain orthogonality throughout the entire symbol period to ensure that the signals can be recovered to the receiving side without distortion.

Different from other modulation formats, the generation and detection process of OFDM is shown in Fig. 5.10. The OFDM transmitter includes QAM mapping, a serial-to-parallel conversion, IDFT, the addition of a cyclic prefix, and a parallel-to-serial conversion. The receiving process is reverse to that of the transmitter. At the transmitter, the data sequence is converted to N parallel symbols after the serial-to-parallel conversion and is then modulated in each subcarrier. Next, the parallel symbols are passed through inverse fast Fourier transform (IFFT) to be a set of N different subcarriers and then added with the guard interval. The generated OFDM signals are subsequently amplified by a power amplifier and then combined with a DC current to drive the LED. At the receiver, the optical signals are detected

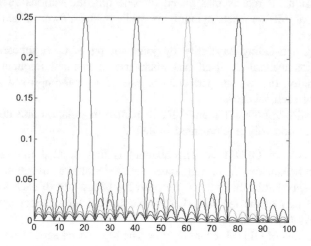

Fig. 5.9 Spectrum of OFDM subcarriers

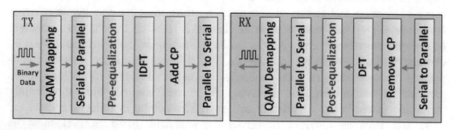

Fig. 5.10 Modulation and demodulation process of OFDM

by a PD and then converted to electrical signals. After OFDM demodulation, the original data can be recovered. Here, the cyclic prefix is utilized to avoid the time delay generated by multipath interference.

Due to the nonflat channel frequency response, the attenuations of the OFDM subcarriers are different. An adaptive modulation method can be used to determine the bit and power loaded onto each subcarrier, according to the SNR of each subcarrier, as shown in Fig. 5.11.

Figure 5.12 shows the constellations of the 16th subcarrier (512QAM) and 512th subcarrier (QPSK).

OFDM modulation technology has been widely used in visible light communications, including both off-line systems and real-time systems. However, OFDM also has some disadvantages. These include large PAPR value and sensitivity to frequency offset. The second issue will be discussed in Chap. 6. Regarding the first question, PAPR value can be reduced by a direct wave cut or PTS technology.

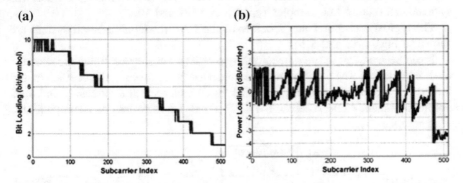

Fig. 5.11 Bit and power loading of OFDM subcarriers

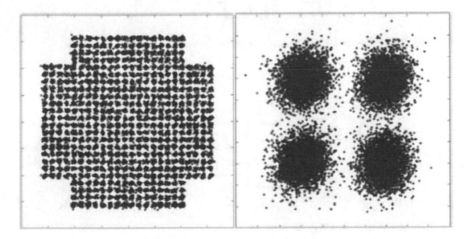

Fig. 5.12 Constellations of different subcarriers

5.5   CAP Modulation Technology

Carrier-less amplitude and phase (CAP) modulation is a type of multidimensional multilevel modulation technology, which was firstly proposed by Bell Lab in the 1970s. CAP modulation allows for relatively high order data using both optical and electrical components of limited bandwidth [14]. By employing a digital filter with several taps, the CAP signal can be generated, and a high-order modulation can be realized [15]. Compared to traditional schemes, such as quadrature amplitude modulation (QAM) and orthogonal frequency division multiplexing (OFDM), the advantage of CAP is that no electrical or optical complex-to-real-value conversion is necessary, therefore it doesn't need a complex mixer and radio frequency (RF) source, or optical IQ modulator. It also does not require the discrete Fourier transformation (DFT) which is utilized in the OFDM signal generation and demodulation; therefore, it can considerably reduce the computation and system structure complexity. We finally conclude that CAP modulation is suitable for systems that require low complexity, such as PON and VLC.

The typical transmitter and receiver of the CAP modulation-based system are shown in Figs. 5.13 and 5.14.

It can be seen that at the transmitter, two orthogonal shaping filters are used for CAP modulation. The CAP signals are generated by controlling the coefficient and the taps of the two shaping filters, which has a better spectral efficiency and does not need a mixer. At the receiver, the CAP signals are recovered by adaptive filters.

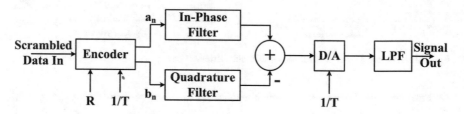

Fig. 5.13 Transmitter of the CAP-based system

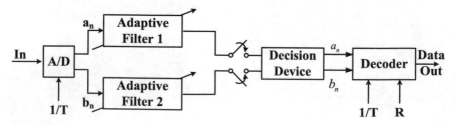

Fig. 5.14 Receiver of the CAP-based system

The signal after CAP modulation can be expressed as:

$$s(t) = \sum_{n=-\infty}^{\infty} [a_n p(t - nT) - b_n \hat{p}(t - nT)] \qquad (5.14)$$

$$p(t) = g(t) \cos(2\pi f_c t), \hat{p}(t) = g(t) \sin(2\pi f_c t) \qquad (5.15)$$

Here, $p(t)$ and $\hat{p}(t)$ are a Hilbert pair, and a_n and b_n are digital signals. In addition, $g(t)$ is the baseband pulse, and both the raised cosine pulse and root-raised cosine pulse are usually used for baseband pulse signals. F_c is the frequency which is set to be larger than the maximum frequency component of $g(t)$.

Assuming $f_c = 1/T$ and T is the symbol time, by using a root-raised cosine pulse for pulse shaping, the time domain and frequency domain filter responses are shown in Figs. 5.15 and 5.16. In addition, the roll-off factors are 0.15, 0.35, and 0.55, respectively.

Due to its simple structure and low computational complexity, CAP modulation shows great potential in visible light communication. Reference [16] has

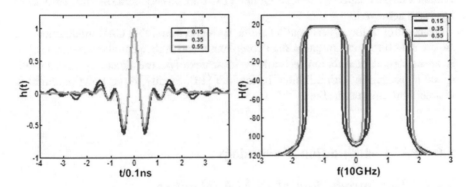

Fig. 5.15 In-phase filter responses in time domain and frequency domain

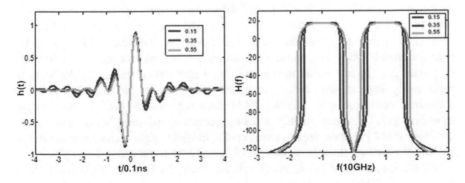

Fig. 5.16 Orthogonal filter responses in the time domain and frequency domain

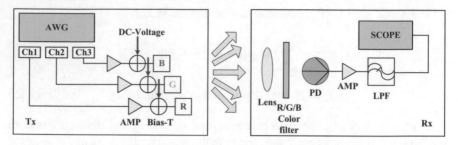

Fig. 5.17 Experimental setup of the CAP modulation-based VLC system

experimentally demonstrated the feasibility of CAP modulation in visible light communication. In Ref. [17], the CAP modulation-based VLC system is shown in Fig. 5.17. In this experiment, a RGB-LED is used as the optical source for the WDM VLC system. Moreover, both pre-equalization and decision feedback equalization are utilized to improve the frequency response of the RGB-LED. In this experiment, CAP signals with different orders are measured to achieve the VLC system's largest capacity. A 3.22 Gbit/s VLC transmission at a distance of 0.25 m has been achieved.

According to the experiment's results, it can be found that CAP modulation has simple structure, low computational complexity, and high spectral efficiency [18]. It is an important modulation technology to achieve spectral efficiency and a high-speed transmission with a limited bandwidth [19]. Finally, it has great potential in visible light communication.

5.6   PAM Modulation Technology

5.6.1   The Introduction of PAM Modulation

PAM (Pulse Amplitude Modulation) method is a kind of one-dimensional multistage modulation technology. Using this modulation technique, high-frequency transmission of high-frequency spectrum can be achieved under the condition of limited bandwidth. Compared with the traditional QAM and OFDM technology, PAM modulation only has real component transformation, and doesn't need to do the plural of electrical or optical signal to the real signal conversion, this makes PAM to be more easily implemented, more convenient and more flexible than QAM. In the meantime, compared with OFDM, PAM does not need to take discrete Fourier transform (DFT), which greatly reduces computational complexity and system structure. PAM has great application potential in visible light communication due to its simple structure, flexible implementation, and low computational complexity.

At the international level, many scholars have studied the high-speed transmission of PAM modulation system. In laser communication, Mohamed morsy-osman et al. used 1310 nm single laser and silicon photon (SiP) intensity modulator to implement the intensity modulation/direct detection of 56-gbaud 4-pam polarization multiplexing (PDM). In the system, the speed of the system is 224 Gb/s and the error rate is 4.1E-3. Mathieu Chagnon et al. used the SiP Mach–Zehnder modulator to implement the 37.4 Gbaud 8-PAM signal in SMF fiber with a speed of 112 Gb/s 10 km transmission, and the BER was lower than 3.8E-3. In visible light communication, Grzegorz Stepniak et al. demonstrated the transmission experiment with speed up to 1.1 Gbit/s in 4-PAM white light VLC system, and the transmission distance was 46 cm.

The advantages and disadvantages of some PAM modulation are summarized below.

Advantages:

- Implementation of digital methods. Various digital signal processing algorithms can be added to improve the performance of the system [20].
- Low complexity and low power consumption. Compared with other technologies such as OFDM, PAM has lower computational complexity, smaller power consumption, simpler structures and is easier to implement.
- High bandwidth utilization. With the increase of the modulating order, the bandwidth decreases.
- The peak-to-average power ratio (PAPR) is lower, so PAM system is less affected by the nonlinear system.

Disadvantages:

- The higher modulation order is more interfered with the higher code, and the precision of the sampling clock is higher.

Solutions:

- The digital equalization algorithm is used to modify the receiving end.
- Use the balanced algorithm on the transmitter to pre-compensate.

5.6.2 The System of PAM-VLC

The VLC system block diagram of PAM modulation technology is shown in Fig. 5.18. The block includes a detailed composition of the PAM launch and receiving end.

At the transmitter, the original data bitstream, which is the random sequence of 0101, is generated. Then, PAM modulation is done. As for PAM-4 modulation, each pair of data bit is coded into a code, and the interval between each code is 2. For example, 00 corresponds to -3, 01 corresponds to -1, 10 corresponds to 1, and 11 corresponds to 3. Corresponding to M-PAM modulation, each $\text{Log}_2 M$ bit is coded into a code, and the corresponding symbol level is $-(M-1) \sim M-1$. The interval of each symbol level is 2. If the symbol level of a signal is M and the desired bit rate is R, then the symbol rate D can be decreased by $\mathrm{Log}_2 M$, that is:

$$D = \frac{R}{\log_2 M} \tag{5.16}$$

After PAM coding, the output signal is sampled for N times; that is, for a data to be represented by the same N data or between adjacent data to insert N minus 1 zeroes, the purpose of the sampling is to realize the n-cycle continuation of the spectrum. The pulse shaping is formed after sampling. The commonly used pulse shaping methods are rectangular pulse shaping, rise cosine pulse shaping (RC), mean square root rise cosine pulse (RRC) shaping, or frequency domain filtering. The signal is then transmitted to the visible channel for transmission. At the receiver, the receiving signal is normalized first, the receiving signal is multiplied by the ratio of average power of the receiving signal and the average power of the transmitting signal.

Then, the normalized receiving signal is sampled at N times, and the post-equalization digital processing algorithm is added to reduce the attenuation and distortion of the signal in the transmission process. Finally, the decoder obtains the original data bit. For M-PAM signal, the $M - 1$ decision threshold is set at the decoding time, which is the average of two adjacent coding levels.

The expressions of PAM modulation signal are shown in formulae (5.17) and (5.18):

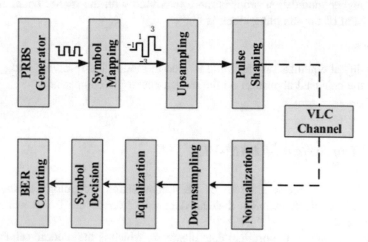

Fig. 5.18 VLC system block diagram of PAM modulation technology

$$s(t) = \sum_{n=0}^{M-1} a_n P(t - nT) \qquad (5.17)$$

$$a_n = -M + 1, -M + 3, \ldots, M - 3, M - 1 \qquad (5.18)$$

where M is the number of coded orders, a_n is the symbol after encoding, and T is the sampling interval.

5.7 Summary

This chapter introduces several modulation technologies suitable for VLC systems. Since LED is a noncoherent light source, only direct modulation techniques can be applied in VLC systems. Both OOK modulation and PPM modulation have unique advantages relating to circuit design and implementation. DMT and OFDM have good performing spectral efficiencies and can overcome multipath interference. In addition, we also introduced the principle of CAP modulation and its corresponding experiments. For high-speed VLC systems, OFDM has become a mainstream technology. In the next chapter, we will describe the signal processing of OFDM.

References

1. Chi, N., Wang, Y., Wang, Y., Huang, X., Lu, C.: Ultra-high speed single RGB LED based visible light communication system utilizing the advanced modulation formats (Invited). Chin. Opt. Lett. **12**(1), 22–25 (2014)
2. Jang, H.J., et al.: PWM-based PPM format for dimming control in visible light communication system. In: International Symposium on Communication Systems, Networks & Digital Signal Processing, pp. 1–5 (2012)
3. Sugiyama, H., Haruyama, S., Nakagawa, M.: Brightness control methods for illumination and visible-light communication systems. In: International Conference on Wireless & Mobile Communications, (IEEE Computer Society, 2007), pp. 78–78
4. Ai-ping, H., et al.: A differential pulse position width modulation for optical wireless communication. In: Conference on Industrial Electronics and Applications, IEEE, pp. 1773–1776 (2009)
5. Sugiyama, H., Nosu, K.: MPPM: a method for improving the band-utilization efficiency in optical PPM. Lightwave Technol. **7**(3), 465–472 (1989)
6. Siddique, A.B., Tahir, M.: Joint brightness control and data transmission for visible light communication systems based on white leds. In: Consumer Communications & Networking Conference, pp. 1026–1030 (2011)
7. Kim, J., Lee, K., Park, H.: Power efficient visible light communication systems under dimming constraint. In: PIMRC, pp. 1968–1973 (2012)
8. Bai, B., Xu, Z., Fan, Y.: Joint LED dimming and high capacity visible light communication by overlapping PPM. In: Wireless & Optical Communications Conference, pp. 1–5 (2010)
9. Wu, Y., Yang, C., Yang, A., et al.: OPPM and PWM concatenated modulation for LED indoor communication. In: Photonics and Optoelectronics. IEEE, pp. 1–3 (2012)

10. Vučić, J., Kottke, C., Nerreter, S., Langer, K., Walewski, J.: 513 Mbit/s visible light communications link based on DMT-modulation of a white LED. J. Lightwave Technol. **28**(24), 3512–3518 (2010)
11. Vučić, J., Kottke, C., Habel, K., Langer, K.: 803 Mbit/s visible light WDM link based on DMT modulation of a single RGB LED luminary. In: Optical Fiber Communication Conference & Explosion, OWB6 (2011)
12. Kottke, C., Hilt, J., Habel, K., Vučić, J., Langer, K.: 1.25 Gbit/s visible light WDM link based on DMT modulation of a single RGB LED luminary. In: Optical Fiber Communication Conference & Explosion, We.3.B.4 (2012)
13. Huang, X., Chen, S., Wang, Z., Shi, J., Wang, Y., Xiao, J., Chi, N.: 2.0-Gb/s visible light link based on adaptive bit allocation OFDM of a single phosphorescent white LED. IEEE Photonics J. **7**(5), 1–8 (2015)
14. Wang, Y., Tao, L., Huang, X., Shi, J., Chi, N.: 8-Gb/s RGBY LED-based WDM VLC system employing high-order CAP modulation and hybrid post equalizer. IEEE Photonics J. **7**(6), 1–7 (2015)
15. Wang, Y., Huang, X., Tao, L., et al.: 4.5-Gb/s RGB-LED based WDM visible light communication system employing CAP modulation and RLS based adaptive equalization. Opt. Express **23**(10), 13626–13633 (2015)
16. Wu, F., et al.: 1.1-Gbps white-LED-based visible light communication employing carrier-less amplitude and phase modulation. IEEE Photonics Technol. Let. **24**(19), 1730–1732 (2012)
17. Wu, F., et al.: 3.22-Gb/s WDM visible light communication of a single RGB LED employing carrier-less amplitude and phase modulation. In: Optical Fiber Communication Conference & Explosion, Th1.G (2013)
18. Wang, Yiguang, et al.: Enhanced performance of a high-speed WDM CAP64 VLC system employing Volterra series-based nonlinear equalizer. IEEE Photonics J. **7**(3), 1–7 (2015)
19. Chi, N., Zhou, Y., Liang, S., Wang, F., Li, J., Wang, Y.: Enabling technologies for high speed visible light communication employing CAP modulation (invited). J. Lightwave Technol. **36**(2), 510–518 (2018)
20. Chi, N., Zhang, M., Zhou, Y., Zhao, J.: 3.375-Gb/s RGB-LED based WDM visible light communication system employing PAM-8 modulation with phase shifted Manchester coding. Opt. Exp. **24**(19): 21663 (2016)

Chapter 6
Visible Light Communication Pre-equalization Technology

In the visible light communication system, the serious unevenness of the channel to the realization of high-speed data transmission has brought great obstacles, although the use of OFDM and other high-order modulation technology to a certain extent optimize system performance, but make the system capacity to further enhance [1]. Transmitter also need to use a variety of pre-equalization technologies, the frequency response to the LED to compensate, thereby improving the system's modulation bandwidth [2, 3]. Pre-equalization technology is divided into two kinds: hardware balance and software balance, the former refers to the use of traditional analog circuits to compensate for signal attenuation; the latter is mainly based on FPGA, that is, field-programmable gate array, designed to meet the requirements of the FIR filter, in order to achieve a balanced effect.

In this chapter, we introduce the hardware pre-equalization circuit and simulate it. At the same time, we introduce software pre-equalization technology, including pre-equalization technology based on FIR filter and software pre-technology based on OFDM modulation. Finally, we introduce ACO-OFDM technology Time domain windowing technology. In the next chapter, we will focus on a series of back-end received signal equalization algorithms, including clock recovery, frequency offset and phase offset estimation, and some back-end processing of OFDM. Pre-equalization combined with equalization will maximize the system of visible light communication systems capacity.

6.1 Hardware Pre-equalization Circuit

In this section, we will use the phosphor LED as an example, the principle and function of the equalizer. We will first introduce an existing hardware-based analog equalization technology: multi-resonant pre-equalization technology introduces its principle and analyzes its performance.

© Tsinghua University Press, Beijing and Springer-Verlag GmbH Germany 2018   111
N. Chi, *LED-Based Visible Light Communications*, Signals and Communication
Technology, https://doi.org/10.1007/978-3-662-56660-2_6

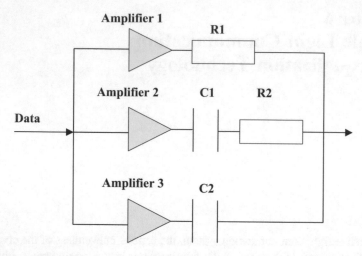

Fig. 6.1 Pre-equalization experiment configuration

With pre-equalization technology, the 3 dB modulation bandwidth of the phosphor LED can be extended from 2 to 25 MHz. By combining the blue filter with the multi-resonant pre-equalization technique, the modulation bandwidth can be further increased to 45 MHz. The transmission rate of 80 Mbit/s can be realized by using NRZ-OOK. The experimental block diagram of this VLC system is shown in Fig. 6.1.

This system uses commercially available phosphor LEDs as VLC transmitters. The LED is driven by a DC current source to achieve the desired brightness. The original data is pre-equalized by drive 1–3 (BUFF634T), and then through Bias-Tee and DC signal combined signal. DC current of 200 mA, while ensuring the lighting conditions, ensures that the device works in a linear area. The VLC receiver consists of a blue filter, a focus lens, a photodetector (PIN type), and a low noise amplifier.

The series inductance introduced by the LED and the driver circuit is 330 nH, and the internal resistance of the LED is 0.9 Ω. The resonant drive technique eliminates the effect of the inductor at the resonant frequency point of $f = 1/2\pi\sqrt{LC}$, maximizing the drive current. The pre-equalizer uses three parallel drivers to equalize the frequency response of the blue component. Amplifier 1, 2, 3 are equalized to the low-frequency, medium-frequency, high-frequency range of the blue frequency response. The equalized bandwidth is determined by the resonant frequency of the high-frequency band, i.e., $f_2 = 1/2\pi\sqrt{LC_2}$, determined by C_2 in the driver 3. In the middle band defined by C_1, the driver 2 contains an additional resistor R_2 to achieve the following purposes: (i) reducing the current flowing to the LED to reduce its higher frequency response compared to the high-frequency band; (ii) the frequency response in the middle band, because the resistance in series with a resistor will reduce the resonant Q value. The frequency response of the low-frequency band corresponds to the frequency response of the middle and high-frequency bands, and the drive current of the driver R1 is regulated by adjustment to achieve equalization.

The researchers measured the frequency response of the LEDs after equalization. With pre-equalization technology, the bandwidth of the phosphor LED can be extended to 45 MHz, above this bandwidth, using NRZ-OOK can achieve 80 Mbit/s data transmission, and BER $<10^{-6}$.

In this section, we will introduce two practical hardware circuits, one is commonly used T-type equalizer and another is improved circuit based on T-type equalizer. Respectively, for its simulation, and for the second circuit, produced the actual PCB circuit board for experimental verification.

6.1.1   Hardware Pre-equalization Circuit Simulation

T-type balanced circuit schematic diagram is shown in Fig. 6.2. In the schematic, role of C5, L2 is to change the high-frequency cutoff frequency and gain amplitude, while controlling the curve of low-frequency gain trend; role of R5 and R2 is to change the lowest frequency of low frequency, that is, the starting range of low frequency.

According to the circuit shown in Fig. 6.2, the appropriate component parameters are selected, and the use of circuit software ADS simulation and the circuit frequency response simulation results are shown in Fig. 6.3. As shown in the graph, the frequency response of the circuit is matched with our expected one.

Based on the T-type equalizer improved circuit shown in Fig. 6.4, the circuit diagram shows that its difference with the typical structure is the use of a series of parallel capacitor resistance to replace the inductance, in particular the use of R3 and C4 parallel circuit to replace the typical T-Type structure in the inductance L2.

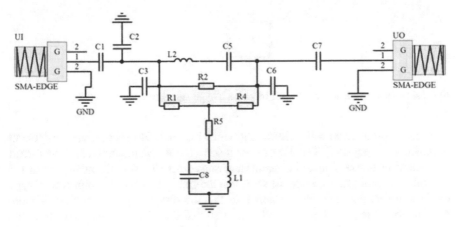

Fig. 6.2 Typical T-type network equalization circuit

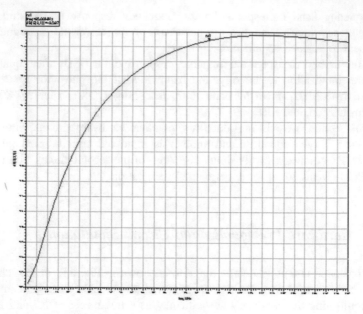

Fig. 6.3 Typical T-type network frequency response curve

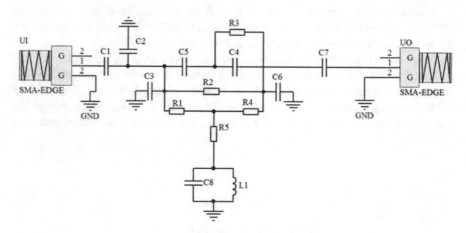

Fig. 6.4 Improved equalization circuit based on T-type equalizer

Also in accordance with the circuit shown in Fig. 6.4, the appropriate component parameters are selected. The use of circuit software ADS simulation and the circuit frequency response simulation results are shown in Fig. 6.5. Compared with the typical structural graphs, it can be seen that the rising trend of the improved T-type equilibrium structure is more steeper in the low-frequency part, which is more suitable than the fast attenuation characteristic of the LED. At the same time, the frequency response does not appear in the larger bandwidth.

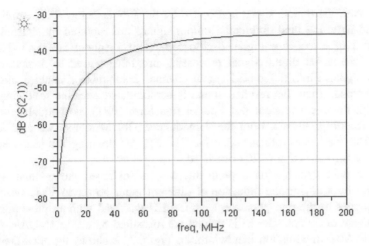

Fig. 6.5 Improved circuit response curve based on T-type equalizer

6.1.2   Hardware Pre-equalization Circuit Experimental Verification

The experimental setup of VLC system is shown in Fig. 6.6, and only the cascaded equalizer is used in the data transmission experiment because of its wider bandwidth. In the experiment, the OFDM signals are generated by a computer using the MATLAB software. The OFDM transmitter consists of QAM modulation, serial-to-parallel conversion, inverse fast Fourier transform (IFFT), cyclic prefix (CP) insertion, parallel-to-serial conversion, up-sample, and up-conversion. Other detailed parameters of generated OFDM signals include: subcarrier number = 128, up-sampling factor = 4. Then, the OFDM signal is supplied to an AWG (Tektronix AWG710), which converts it to an analog signal as the input of the VLC system. At

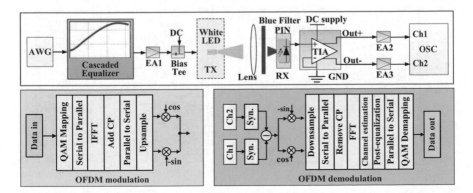

Fig. 6.6 Experimental setup of VLC system

the receiver, the optical signal is detected by the PIN detector. After amplified by TIA and EAs, the final differential output signals are captured by channel 1 and channel 2 of a real-time digital oscilloscope (OSC, Agilent 54855A). Then, an off-line MATLAB digital signal processing program is used to demodulate the OFDM signal. First, differential signal channel 1 subtracts channel 2 after synchronizations. Then, the resulted signal is demodulated after the following steps, including down-conversion, fast Fourier transform (FFT), post-equalization, and QAM decoding. At last, from the demodulated OFDM signal, we calculate the BER. The total data rate includes the 3% CP, 5% training sequence, and 7% forward error correction (FEC) overhead.

In the data transmission experiment, we first obtained the optimal working condition by studying the influence of different input power to VLC system and bias voltages of the phosphorescent white LED with and without pre-equalizer at the distance of 50 cm. The BER results are measured based on 16-QAM-OFDM with 300 MHz modulation bandwidth. In Fig. 6.7, it shows the measured BER results versus different input power from −40 to −17 dBm with pre-equalizer and without equalizer. The electrical power is measured after the output of the AWG by spectrum analyzer HP 8562A and controlled by varying the signal driving peak-to-peak voltage (Vpp) of AWG. The bias voltage of the phosphorescent white LED is fixed at 3.3 V. The optimal input powers −18.4 and −34.0 dBm of system with pre-equalizer and without pre-equalizer are obtained, and the BER results are, respectively, 4.24×10^{-3} and 6.64×10^{-4}. The constellations (i) and (iii) in Fig. 6.7 are systems with and without pre-equalizer at −34.0 dBm, while (ii) and (iv) are without and with pre-equalizer at −18.4 dBm. The system with equalizer has better BER performance when compared with the optimal BER performance. Using the equalizer, the BER results of VLC system can be greatly reduced by an order of magnitude. The optimal input power to VLC system with pre-equalizer is higher than the one without pre-equalizer, so the systems with equalizer need more transmission power to achieve the best working condition. In Fig. 6.8, we show the BER results versus bias voltage from 2.9 to 3.5 V with equalizer and without

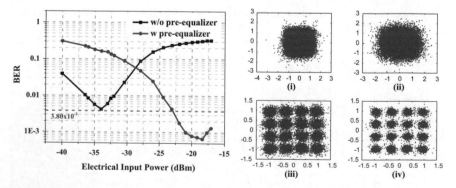

Fig. 6.7 Measured BER results versus different input power to VLC system with pre-equalizer and without equalizer

equalizer, and the input power is fixed at −18.4 and −34.0 dBm, respectively. The bias voltage of white LED at 3.2 V is the optimal point of system with and without pre-equalizer systems, and the corresponding BER results are 5.27×10^{-4} and 3.94×10^{-3} with constellations (v) and (vi) inserted in Fig. 6.9, respectively.

At the optimal input power −18.4 dBm and bias voltage 3.2 V, we measured the BER results versus different bandwidths from 200 to 450 MHz with 50 MHz step, different modulation orders from QPSK to 32-QAM, and different distances from 50 to 200 cm with 50 cm step using the pre-equalizer, and distances at 100 and 200 cm are presented in Fig. 6.9. The total data rate can be calculated as $R = B*\log 2(M)$, where B is the modulation bandwidth of the system and M is the constellation size of QAM. From Fig. 6.9, the BER performance degrades with the increase of modulation bandwidth, modulation order, and transmission distance. The high frequency after about 300 MHz fades fast, resulting to lower SNR in this part. Higher modulation orders require higher SNR to achieve the FEC limit. As the

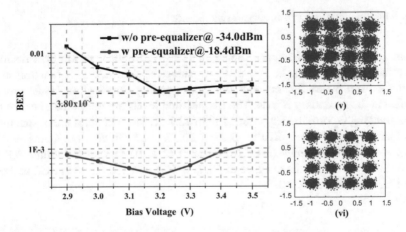

Fig. 6.8 Measured BER results versus bias current with pre-equalizer @−18.4 dBm and without equalizer @−34.0 dBm

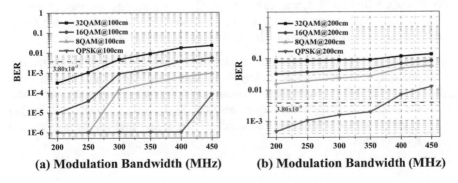

Fig. 6.9 BERs versus various modulation bandwidths **a** distance = 100 cm; **b** distance = 200 cm

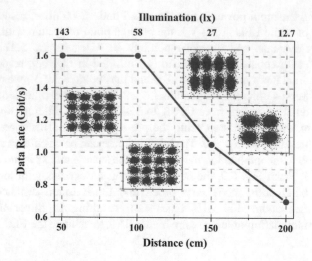

Fig. 6.10 Total data rate versus transmission distance

transmission distance becomes longer, the received optical power will decrease, especially after 100 cm. In Fig. 6.10, we show the data rate as a function of the distance from 50 to 200 cm and the corresponding illumination levels are also plotted in the secondary X axis. The illumination levels are measured after the blue filter, ranging (approximately) from 143 to 12.7 lx. The data rates are, respectively, 1.6 Gbit/s (16-QAM, BER: 3.27×10^{-3}), 1.6 Gbit/s (16-QAM, BER: 3.62×10^{-3}), 1.05 Gbit/s (8-QAM, BER: 3.32×10^{-3}), and 0.7 Gbit/s (QPSK, BER: 1.93×10^{-3}) at the distance of 50, 100, 150, and 200 cm with BER under the pre-FEC limit of 3.8×10^{-3}.

6.2   Software Pre-equalization

As described in Sect. 6.1, the existing equalization schemes are mostly implemented using conventional analog circuits, although they can increase the bandwidth of the system to a certain extent, there are still a lot of limitations:

(1) analog circuits exist time fluter, weak anti-interference, bandwidth limitations, and other shortcomings do not apply to high-speed signal transmission.
(2) lack of flexibility. Visible light channel is affected by environmental noise, and the flexibility of equalizer adjustment has higher requirements. The analog circuit cannot be based on the needs of the actual channel debugging and improvement.

Software equalization technology, which can be flexibly adjusted according to the system requirements, has a series of advantages. In this section, we have studied the software-based equalization technology, including pre-equalization based on FIR filter and pre-equalization based on OFDM modulation technology.

6.2.1 Pre-equalization Technology Based on FIR Filter

In this section, the principle of FIR filter is introduced first; then the principle of time domain equalizer is introduced. Finally, the design method of FIR filter in VLC system is introduced. It is hoped that the reader will have a whole understanding of software equalization method.

6.2.1.1 Principle of FIR Filter

FIR (Finite Impulse Response) filter: A finite unit of impulse response filter is the most basic component of a digital number system. The equation of the input and the output of FIR system is as followed:

$$y[n] = \sum_{k=0}^{N-1} h(k)x(n-k) \tag{6.1}$$

So the FIR filter system function is:

$$H(z) = \frac{Y(z)}{X(z)} = \sum_{n=0}^{N-1} h(n)z^{-n} \tag{6.2}$$

Since the unit impulse response $h(n)$ of the FIR filter is a finite length sequence, $H(z)$ is the $(N-1)$ polynomial of Z^{-1}, which has $(N-1)$ zero points on the Z plane, $(N-1)$ order heavy pole. Therefore, $H(z)$ is always stable. The FIR filter design task is to select a finite length of $h(n)$, so that the transfer function $H(e^{j\omega})$ satisfies certain amplitude characteristics and linear phase requirements. Since the FIR filter is easy to implement a strictly linear phase, the core idea of the FIR digital filter design is to find a finite impulse response to approximate the given frequency response.

There are two main methods of designing FIR filter. One is window function method, and another is frequency sampling method. The process of designing a FIR filter using the window function method is as follows:

Step 1: Given the filter frequency response function $H(e^{j\omega})$ to be designed.
Step 2: Find the filter impulse response $h(n) = \text{IFFT}[H(e^{j\omega})]$.

Step 3: Select the window function $w(n)$ and the window size (i.e., filter order) N, and commonly used window function such as rectangular window, angle window, Hamming window, Hanning window, Kaiser window.

Step 4: Find the unit sampling response $h_d(n) = h(n)w(n)$ of the designed FIR filter.

Step 5: Get the FIR filter frequency response function $H_d(e^{j\omega}) = \text{FFT}[h(n)]$, and test whether to meet the design requirements.

In summary, the FIR filter has the following advantages: (1) arbitrary amplitude-frequency characteristics, (2) strict linear phase, (3) unit sampling response is limited, so the filter is a stable system, (4) achieved with a causal system (because as long as after a certain delay, any non-causal finite length sequence can become a causal finite length sequence), (5) impulse response is limited, you can use FFT frequency offset compensation algorithm, to improve the efficiency of the operation, (6) to avoid similar to the analog filter time jitter. Because of the many advantages of FIR filter, it has a wide range of applications in communication, image processing, pattern recognition and other fields.

6.2.1.2   Time Domain Equalizer

Equilibrium can be divided into frequency domain equalization and time domain equalization. The so-called frequency domain equalization, from the correction system frequency characteristics, includes the equalizer, the baseband system, and the total characteristics of non-distortion to meet the transmission conditions; the so-called time domain equalization is the use of equalizer that generated time response to correct distortion. The impulse response of the entire system, including the equalizer, satisfies the inter-coded crosstalk condition.

Frequency domain equalization is not true for channel characteristics and is applicable when transmitting low-speed data. The time domain equalization can be adjusted according to the change of channel characteristics, which can effectively reduce the inter-symbol crosstalk, so it can be widely used in high-speed data transmission.

As shown in Fig. 6.11, here is a kind of time domain equalizer. The network consists of an infinite number of horizontally arranged delay cells and tap coefficients, so it is called a transversal filter.

The function of a transversal filter is to transform the response waveform with inter-symbol interference, with the sampling time at the inputs (outputs of the receiving filter), to the response waveform without inter-symbol interference. The equilibrium principle of a transversal filter is built on a response waveform, so it is called time domain equalization.

From the above analysis, we can see that the transversal filter can realize time domain equalization. In theory, an infinitely long transversal filter can completely eliminate inter-symbol interference during the sampling time, but in fact, this is

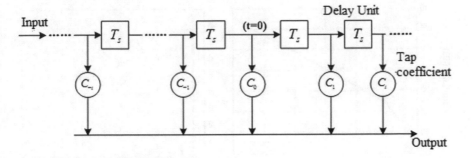

Fig. 6.11 Transversal filter structure

impossible. This is because the length of the equalizer is not only restricted by economic conditions, but it is also restricted by the adjustment accuracy of every factor, C_i. If the adjustment accuracy of C_i cannot be guaranteed, even if the length is increased, the effect will not be significant. The designed principle based on a FIR filter not only realizes a limited long transversal filter, but can also determine the coefficient C_i of each tap, which strengthens the practicality of the transversal filter.

6.2.1.3 The Design of a Pre-equalizer Based on a FIR Filter

Figure 6.12 shows the application of a pre-equalizer based on a FIR filter that is designed for the VLC system.

After the frequency response curve analysis of white LEDs in the third chapter, it can be concluded that the frequency response characteristic curve of an ideal filter is shown in Fig. 6.13a. Its segment slope is:

$$s = \begin{cases} 1.02 \text{ dB/MHz}, \ 0 \le \omega \le 10 \text{ MHz} \\ 0.42 \text{ dB/MHz}, \ 10 \text{ MHz} \le \omega \le 60 \text{ MHz} \end{cases} \tag{6.3}$$

First, make the frequency response curve discretization, and then do the IFFT with MATLAB software. The corresponding time domain impulse response is shown in Fig. 6.13b.

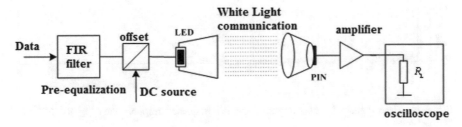

Fig. 6.12 A system block diagram of a VLC system using a FIR filter as pre-equalizer

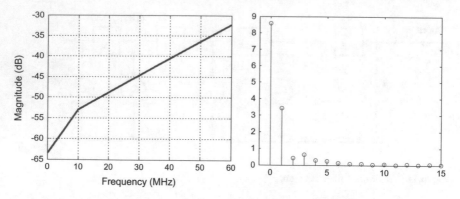

Fig. 6.13 Frequency response curve and the time domain impulse response of an ideal filter

The design principle of a FIR filter is to truncate the time domain impulse response in the Fig. 6.13b by adding a window, so as to obtain an approximate ideal filter frequency response curve. This example uses the Kaiser window function. Figures 6.14, 6.15, 6.16, and 6.17 show the time domain impulse responses and the frequency response curves of the FIR filter with the order number ranging from 2 to 5. Due to the effective values of the time domain impulse response, an ideal filter mainly focus on $N = 0$ and $N = 1$, so the 2-order FIR filter is generally excluded from being considered as an ideal filter. Of course, the higher the order number is, the closer the FIR filter design will be to the ideal filter. However, the filter order is proportional to the complexity of the module, so in the actual design process, we need to balance the relationship between the filter's performance and complexity by selecting the appropriate filter order.

Use the window function method to design the FIR filter according to Eq. 6.3, where the window function used in the simulation is the Kaiser window. Without

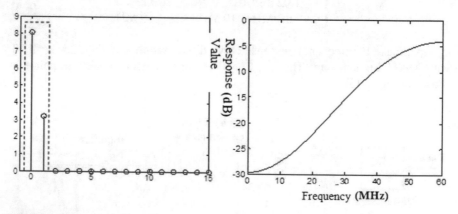

Fig. 6.14 Time domain impulse response and the frequency response curve of the FIR filter @order = 2

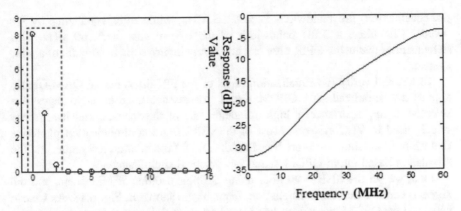

Fig. 6.15 Time domain impulse response and the frequency response curve of the FIR filter @order = 3

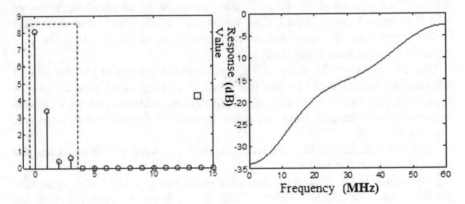

Fig. 6.16 Time domain impulse response and the frequency response curve of the FIR filter @order = 4

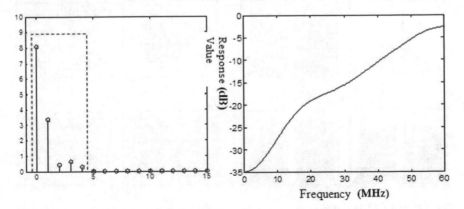

Fig. 6.17 Time domain impulse response and the frequency response curve of the FIR filter @order = 5

pre-equalization, the bandwidth is just 2.5 MHz, while after equalizing by the 4-order FIR filter, a 3 dB bandwidth of the system can reach 60 MHz. This demonstrates that using a FIR filter can effectively improve the bandwidth of a VLC system.

To test and verify the equalization effect of the FIR filter, the 16-QAM-OFDM signals are simulated. An OFDM signal possesses numerous advantages like selective fading resistance, a high utilization rate of the carriers, and it has been widely used in VLC systems. Most of the visible light communication platforms that are built in laboratories are based on the OFDM modulation technique, and the simulation based on an OFDM signal has practical significance.

First, we set the SNR = 20 after using a different order of FIR filters, and the signal spectrums and the constellation diagrams are shown in Fig. 6.18. As we can see, even the OFDM technology has a good selective fading resistance, but because the frequency response curve attenuation of the white LED is very serious (negative exponential attenuation), the constellation diagram of the received signal is very fuzzy. After the equalization of the FIR filter, the constellation points can converge near the standard points. In addition, the increase of the filter order makes the OFDM signal spectrum more flat, so the convergence of the points in the constellation diagram is also improved.

Figure 6.19 shows the study of BER performance and EVM performance of different filter orders. It can be seen that for an OFDM signal, with the increase of the filter order, the noise tolerance of the system also increases and the anti-noise performance is enhanced. Here, the noise tolerance of five-order FIR filters can reach up to 15 dB.

Next, we will explore the connection between the order of FIR filters and the signal modulation format.

As shown in Fig. 6.20, it is the constellation diagram of a 512-QAM signal after the equalization of different orders of FIR filters. It can be seen that with the

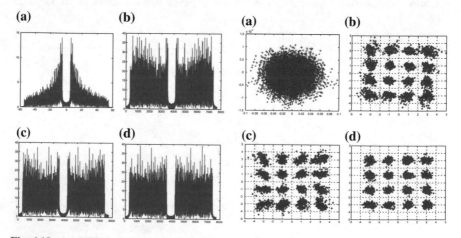

Fig. 6.18 An OFDM signal spectrum and the constellation diagram **a** without equalization and after **b** 2-order **c** 3-order **d** 4-order FIR filter equalization @ SNR = 20

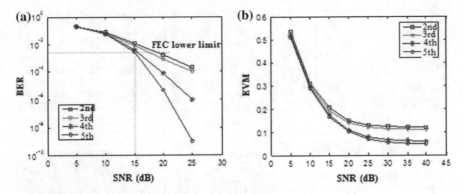

Fig. 6.19 **a** BER (bit error rate) and **b** the EVM (error vector magnitude) of the 16-QAM-OFDM signal with the change of the SNR

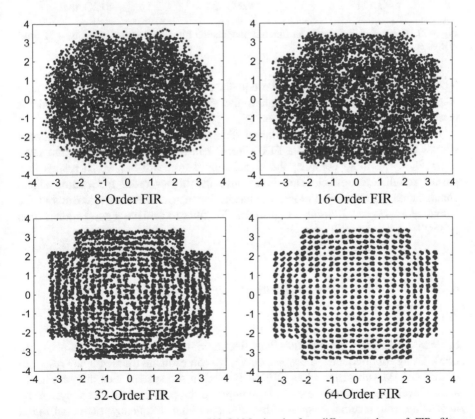

Fig. 6.20 Constellation diagram of a 512-QAM signal after different orders of FIR filter equalization

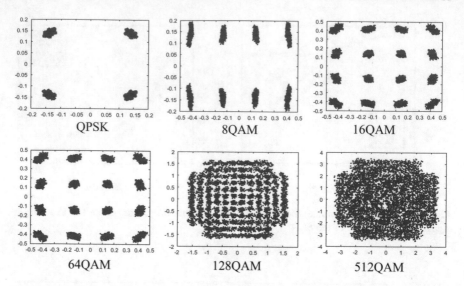

Fig. 6.21 Constellation diagram for different modulation format signals equalized by a 32-order FIR filter

increase of the FIR filter order, the equalization effect is better. This is because when the FIR filter order is higher, the frequency response characteristic is increasingly closer to the ideal filter.

Figure 6.21 shows the constellation diagram for different modulation format signals equalized by a 32-order FIR filter. It can be seen that by using the same order filter, the more complex the signal modulation will be, which results in a poorer equalization effect. This is because the higher order modulation signal constellation points have a closer Euclidean distance, and the requirements of the equalizer are higher. It needs a high-order FIR filter to achieve a good equalization effect.

6.2.2 Software Pre-equalization Technology Based on OFDM

In Sect. 6.2.1.3, we get the blue-light frequency response curve of the phosphor LED. Using this frequency response curve, we can estimate the frequency response of any OFDM subcarrier when OFDM is used in the system. For different attenuation levels subcarriers, we can easily improve the performance of the system by improving the channel response by corresponding frequency pre-equalization techniques. Specifically, we can use the pre-equalization technique to increase the high-frequency signal gain or reduce the low-frequency signal gain, resulting in a flat channel response.

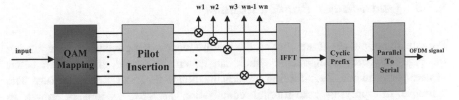

Fig. 6.22 Schematic diagram of frequency pre-equalization based on OFDM modulation

A block diagram of frequency pre-equalization based on OFDM modulation is shown in Fig. 6.22. It consists of QAM mapping, pilot insertion, IFFT transform, join cyclic prefix and parallel-to-serial conversion. Before the IFFT transform, the equalizer is added, and the subcarrier amplitude is equalized by multiplying different OFDM subcarriers by different equalization coefficients. The OFDM signal contains N subcarriers, expressed as $X = (x_1, x_2, \ldots, x_N)$, the received frequency domain signal is denoted as $Y = (y_1, y_2, \ldots, y_N)$, and the equalization coefficient of each subcarrier is denoted by $W = (w_1, w_2, \ldots, w_N)$, and the relationship between the three is given by Eq. (6.4). Therefore, after obtaining the frequency domain information of the signal arriving at the receiving end, we can calculate the equalization coefficient as shown in Eq. (6.5). After calculating the equilibrium matrix, multiply the subcarriers by the corresponding equalization coefficients, and we can get the equalized signal, as shown in Eq. (6.6). Using this equalization technique, a flat subcarrier frequency response can be obtained and the unbalanced gain in the subcarrier band cannot be overcome, but the performance of the system can be significantly improved [4].

$$
\begin{pmatrix} x_1 \\ x_2 \\ \vdots \\ x_l \\ \vdots \\ x_N \end{pmatrix} = \begin{pmatrix} y_1 & 0 & \cdots & 0 \\ 0 & y_2 & \cdots & 0 \\ \vdots & \vdots & \ddots & \vdots \\ 0 & 0 & \cdots & y_N \end{pmatrix} = \begin{pmatrix} w_1 \\ w_2 \\ \vdots \\ w_l \\ \vdots \\ w_N \end{pmatrix} \tag{6.4}
$$

$$
W = \begin{pmatrix} w_1 \\ w_2 \\ \vdots \\ w_l \\ \vdots \\ w_N \end{pmatrix} = \begin{pmatrix} y_1 & 0 & \cdots & 0 \\ 0 & y_2 & \cdots & 0 \\ \vdots & \vdots & \ddots & \vdots \\ 0 & 0 & \cdots & y_N \end{pmatrix}^{-1} \begin{pmatrix} x_1 \\ x_2 \\ \vdots \\ x_l \\ \vdots \\ x_N \end{pmatrix} = \begin{pmatrix} \frac{x_1}{y_1} \\ \frac{x_2}{y_2} \\ \vdots \\ \frac{x_N}{y_N} \end{pmatrix} \tag{6.5}
$$

$$
X_{\text{pre}} = (w_1 x_1, w_2 x_2, \ldots, w_l x_l, \ldots, w_N x_N) \tag{6.6}
$$

6.2.3 Quasi-linear Pre-equalization

Pre-equalization aims at attenuating the low-frequency components and amplifying the high-frequency components which can flatten the whole spectrum for better system performance. Zero-forcing pre-equalization is the most traditional way. The principle of applying zero-forcing equalization in VLC system is shown in Fig. 6.23. Although zero-forcing pre-equalization can lead to the most flatten received spectrum theoretically, it is not the most suitable software pre-equalization method when applying bit and power loading OFDM in a practical VLC experimental system. The experimental equipments in VLC system cause the limitation. Arbitrary waveform generator (AWG) has output peak power limitation, so the magnitude of output signal should not be too high. The transmitting signal is loaded into a phosphorescent white light-emitting diode (LED), and the signal should belong to the linear work region of the LED. Electrical amplifier (EA) also has output peak power limitation. Besides, if the receiving signal magnitude exceeds PIN receivers' dynamic range, PIN receivers' cannot detect the receiving signal linearly. Thus, the experimental VLC system is a total power and peak power constrained system. Applying bit and power loading OFDM becomes a constraint problem of rate maximization (RM). Zero-forcing pre-equalization is not suitable anymore because the experimental VLC system is not ideal.

Considering the limitation, we proposed three kinds of quasi-linear software equalization lines shown in Fig. 6.24, including linear kind, concave kind, and convex kind. This system uses 512 subcarriers [5]. Concave kind and convex kind both have oblique line part and linear line part, k is used to denote the slope of the oblique line part. k is also used to denote the slope of the linear kind. To test the system performance employing different software pre-equalization lines, we carried out simulation and experiment demonstration. The simulation flowchart is shown in Fig. 6.25. To simplify the problem, we only consider the peak power limitation of EA and the dynamic range if PIN receivers. If the transmitting signals exceed EA peak power limitation or PIN receivers' dynamic range, k of the pre-equalization lines will be adjusted automatically to find the most suitable k for the maximum transmission data rate. The whole simulation system is then finished with the most suitable parameters for the experimental system.

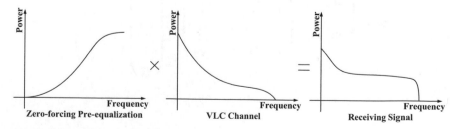

Fig. 6.23 Zero-forcing pre-equalization

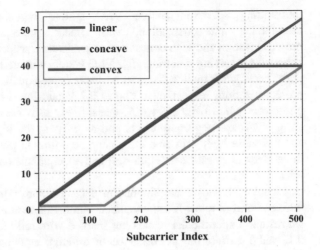

Fig. 6.24 Quasi-linear software equalizer lines

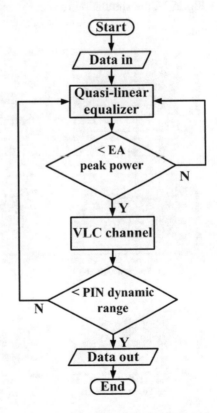

Fig. 6.25 Simulation flow chart

Both simulation and experimental demonstration are carried out to test and verify the performance of different kind of quasi-linear pre-equalization. Measured spectra (Spectrum analyzer, HP8562A) are in Fig. 6.26a, b, d, f, h showing the transmitting spectrum before VLC system without pre-equalization, with zero-forcing pre-equalization, with linear pre-equalization, with convex pre-equalization and with concave pre-equalization, respectively. Figure 6.26c, e, g, i show the receiving spectrum after VLC system with zero-forcing pre-equalization, with linear pre-equalization, with convex pre-equalization and with concave pre-equalization, respectively. Vpp is 0.7 V and the driving current of the phosphorescent white LED is 94 mA. Quasi-linear software equalizer achieves attenuating low-frequency components and amplifying the high-frequency components and has more high gain area than using zero-forcing pre-equalization. Among the three kinds of quasi-linear equalizer, convex equalization has the largest high gain area.

Figure 6.27 shows the simulation data rate and experimental data rate using different kind of quasi-linear equalizer. Simulation results are shown using colorful curves and experimental results are marked with red blocks. We select $k = 0.01$, 0.1, and 1 to test the performance of different equalization lines. According to Fig. 6.27, the simulation results and experimental results are matched. Zero-forcing

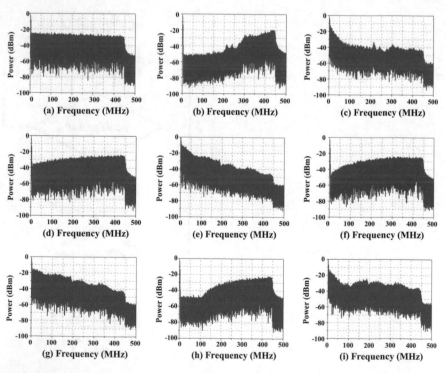

Fig. 6.26 Measured electrical spectra before VLC system **a** without pre-equalization, **b** with zero-forcing pre-equalization, **d** with linear pre-equalization, **f** with convex pre-equalization, **h** with concave pre-equalization and receiving electrical spectra after VLC system, **c** with zero-forcing equalization, **e** with linear pre-equalization, **g** with convex pre-equalization, **i** with concave pre-equalization

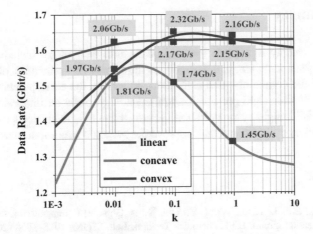

Fig. 6.27 Simulation and experimental data rate using different kind of quasi-linear equalizer

equalization only achieves 0.5-Gb/s data rate. As k increases, the performance of the concave kind becomes better and then worse. When k is small, the performance of the linear kind is better than the other two kinds. But as k increases, the performance of the convex kind approaches the linear kind. The convex kind implements the highest experimental data rate 2.32-Gb/s over 1 m distance when $k = 0.1$ with BER lower than 3.8×10^{-3}. To our knowledge, this is the highest transmission data rate based on a phosphorescent white LED VLC system. The results also verify that zero-forcing pre-equalization is not suitable for experimental VLC systems because of the limitation caused by experimental equipments.

The bit allocation using the convex kind pre-equalization when $k = 0.1$ is shown in Fig. 6.28. There is a mapping between bit number and M-ary QAM, such as 1 bit stands for BPSK, two bits stand for QPSK, three bits stand for 8-QAM, four bits stand for 16-QAM. The corresponding M-ary QAM system constellations are also shown in Fig. 6.28. The total bit error rate of the system is 3.7×10^{-3} under the limitation of forward error correction.

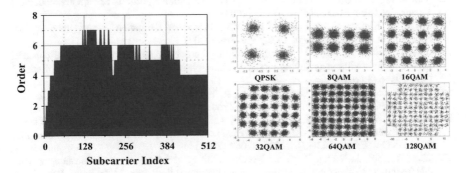

Fig. 6.28 Bit allocation for 512 subcarriers applying the convex kind equalization when $k = 0.1$

6.3 Summary

This chapter mainly introduces the pre-equalization technology including hardware pre-equation and software post-equation, it introduces the principles of these technologies and the commonly used algorithms, as well as the corresponding supporting experiments. Visible light pre-equalization technology plays an important role in the high-speed visible light communication system. It can effectively reduce the crosstalk between subcarrier signals and thus improve the system's maximum transmission rate.

References

1. Huang, X., Shi, J., Li, J., Wang, Y., Chi, N.: A Gbps VLC transmission using hardware pre-equalization circuit. IEEE Photonics Technol. Lett. 27(18), 1915–1918 (2015)
2. Huang, X., Chen, S., Wang, Z., Wang, Y., Chi, N.: 1.2 Gbit/s visible light transmission based on orthogonal frequency-division multiplexing using a phosphorescent white light-emitting diode and a pre-equalization circuit. Chin. Opt. Lett. 13(10), 100602 (2015)
3. Huang, X., Wang, z., Shi, J., Wang, Y., Chi, N.: 1.6 Gbit/s phosphorescent white LED based VLC transmission using a cascaded pre-equalization circuit and a differential outputs PIN receiver. Opt. Exp. J. 23(17), 22034–22042 (2015)
4. Wang, Y., Tao, L., Wang, Y., Chi, N.: High speed WDM VLC system based on multi-band CAP64 with weighted pre-equalization and modified CMMA based post-equalization. IEEE Commun. Lett. 18(10), 1719–1722 (2014)
5. Wang, Y., Chi, N., Wang, Y., et al.: High-speed quasi-balanced detection OFDM in visible light communication. Opt. Exp. 21(23), 27558–27564 (2013)
6. Cherry, S.: Edholm's law of bandwidth. IEEE Spectr. 41(7), 58–60 (2004)
7. Siddique, A.B., Tahir, M.: Joint brightness control and data transmission for visible light communication systems based on white LEDS. CCNC, 1026–1030 (2011)
8. Le Minh, H., Ghassemlooy, Z., O'Brien, D., Faulkner, G.: Indoor gigabit optical wireless communications: challenges and possibilities. ICTON, 1–6 (2011)
9. Le Minh, H., O'Brien, D., Faulkner, G., et al.: 80 Mbit/s visible light communications using pre-equalized white LED. In: ECOC 2008, Brussels, Belgium, P.6.09, 21–25 Sept (2008)
10. Chunmiao, L.: Research on OFDM broadband 4 shortwave technology. Master thesis of Zhejiang University (2011) (in Chinese)
11. Gini, F., Giannakis, G.B.: Frequency offset and symbol timing recovery in flat-fading channels: a cyclostationary approach. IEEE Trans. Commun. 46(10), 400–411 (1998)
12. Yu, Z., Shi, Y., Su, W.: A blind carrier frequency estimation algorithm for digitally modulated signals. In: IEEE MILCOM'04, 48–53 (2004)
13. Leven, A., Kaneda, N., Koc, U.V., Chen, Y.K.: Frequency estimation in intradyne reception. IEEE Photonics Technol. Lett. 19(6), 366–368 (2007)
14. Cao, Y., Song, Y., Shen, J., Wanyi, G., Ji, Y.: Frequency estimation for optical coherent MPSK system without removing modulated data phase. IEEE Photonics Technol. Lett. 22(10), 691–693 (2010)
15. Cao, Y., Yu, S., Chen, Y., Gao, Y., Gu, E., Ji, Y.: Modified frequency and phase estimation for M-QAM optical coherent detection. In: Proc ECOC 2010, 19–23 (2010)
16. Huang, X., Wang, Z., Shi, J., Wang, Y., Chi, N.: 1.6 Gbit/s phosphorescent white LED based VLC transmission using a cascaded pre-equalization circuit and a differential outputs PIN receiver. Opt. Express 23(17), 22034–22042 (2015)
17. Zhou, Y., Shi, J., Wang, Z., et al.: Maximization of visible light communication capacity employing quasi-linear pre-equalization with peak power limitation. Math. Probl. Eng., 2016, (2016-7-25), 2016, 1–8 (2016)

Chapter 7
Visible Light Communication Post-equalization Technology

7.1 Time Domain Equalization Technique

Equalization can be divided into two categories: frequency domain equalization and time domain equalization. The so-called frequency domain equalization corrects the system's frequency characteristics, so that the total characteristics, including the baseband system equalizer, satisfy the undistorted transmission conditions. The so-called time domain equalization directly corrects distorted waveforms to satisfy the requirements of no inter-symbol interference criterion in the time domain, with parameters generated by the equalizer.

Frequency domain equalization is applicable in both the stable channel and in low-speed data transmission. Time domain equalization can be adjusted according to the change in channel characteristics, as well as effectively reduce inter-symbol interference, so it can be widely applied in high-speed data transmissions.

The current mainstream of time domain equalization algorithm is based on the finite length unit impulse response filter (FIR filter) to design, so this section will be the mainstream of several advanced time domain equalization algorithms such as CMA, CMMA, M-CMMA and DD-LMS which are introduced in detail.

7.1.1 CMA Algorithm

Constant modulus algorithm (CMA) is first proposed by Godard based on Bussgang type of blind equalization algorithm and is one of the most commonly used kinds. CMA is a special case of Godard algorithm, by using the gradient descent method to adjust the tap of the equalizer coefficients in order to reduce the cost function. It has great advantages as for low computing complexity and is easy to implement, so become widely used in communication system. The CMA algorithm is applicable to the equalization of all constant envelope signals and part

© Tsinghua University Press, Beijing and Springer-Verlag GmbH Germany 2018
N. Chi, *LED-Based Visible Light Communications*, Signals and Communication
Technology, https://doi.org/10.1007/978-3-662-56660-2_7

of unsteady envelope signals. In coherent communication systems, the CMA algorithm is mainly used in digital coherent system of polarization mode dispersion (PMD) compensation; CMA blind equalization algorithm does not need a known training sequence and is not restricted to limited signal distortion and deformation. This is a high-efficient, excellent blind equalization algorithm of low computational complexity. What's more, its cost function is a concave curve, so at the same time it can be a time domain equalization algorithm to compensate the ISI. CMA algorithm processes are shown in Fig. 7.1.

For such CMA transversal equalizer, the equalizer output can be expressed as

$$y(n) = X^T(n)w(n) \qquad (7.1)$$

In this way, the cost function of CMA algorithm can be expressed as

$$D^{(p)} = E \cdot (|y(n)|^p - R_p)^2 \qquad (7.2)$$

In the formula, R_p is positive real constant. For this kind of cost function, the transmitted signal power can be constant, and the output signal power of the equalization also can be constant.

The minimum value of $D^{(p)}$ can be derived in recursive in accordance with the steepest descent method:

$$w(n+1) = w(n) - \mu_p \frac{\partial D^{(p)}}{\partial w(n)} \qquad (7.3)$$

In formula (7.3), μ_p is the step; generally, the value is small enough. From formula (7.1), we know

$$\frac{\partial |y(n)|}{\partial w(n)} = \frac{\partial}{\partial w(n)} |x^T(n)w(n)|$$
$$= x^*(n)x^T(n)w(n)|x^T(n)w(n)|^{-1} \qquad (7.4)$$

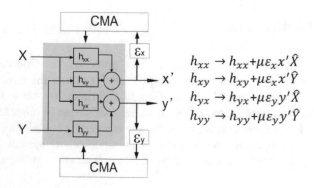

Fig. 7.1 CMA algorithm processes

From formulas (7.4) and (7.2), we can obtain

$$\frac{\partial D^{(p)}}{\partial w(n)} = 2pE[x^*(n)x^{\mathrm{T}}(n)w(n)|x^{\mathrm{T}}(n)w(n)|^{p-2}(|x^{\mathrm{T}}(n)w(n)|^p - R_p)]$$
$$= 2pE[x^*(n)y(n)|y(n)|^{p-2}(|y(n)|^p - R_p)] \tag{7.5}$$

And then putting formula (7.5) into formula (7.3), we get

$$w(n+1) = w(n) + \mu_p 2pE[x^*(n)y(n)|y(n)|^{p-2}(|y(n)|^p - R_p)]$$
$$= w(n) + \mu E[x^*(n)y(n)|y(n)|^{p-2}(|y(n)|^p - R_p)] \tag{7.6}$$

Using instantaneous estimates $x^*(n)y(n)|y(n)|^{p-2}(y(n) - R_p)$ in place of the statistical average in formula (7.6), we can get

$$w(n+1) = w(n) + \mu x^*(n)y(n)|y(n)|^{p-2}(|y(n)|^p - R_p) \tag{7.7}$$

When the equalizer completely achieves balance, its output shall be

$$y(n) = d(n)e^{j(\psi + 2\pi\Delta f n T_s)} \tag{7.8}$$

In formula (7.8), $d(n)$ is the transmitted signal, ψ is the fixed delay, and Δf is the frequency variations. The signal that equalizer received can be written as

$$x(t) - \sum_n d(n)h(t - nT_s)e^{j\varphi(t)} + n(t) \tag{7.9}$$

where $h(t)$ expresses the equivalent unit impulse response from the transmitter to the receiver end channel and $n(t)$ is the additive white Gaussian noise.

The aforementioned formulas (7.1), (7.7), and (7.10) constitute the complete CMA algorithm, especially, generally we let $p = 2$, and then we can get

$$R_p = \frac{E|d(n)|^{2p}}{E|d(n)|^p} \tag{7.10}$$

The corresponding feedback error usually can be expressed as

$$\varepsilon(n) = y(n) - R_2 \tag{7.11}$$

It can be seen that the value of R_2 does not change with received signals; namely, it is not affected by the channel. CMA algorithm has low computing complexity, fast convergence speed, and stable performance, and it can make the closed eye diagram open and makes the fuzzy constellation points clear, so it has been widely used. Its related information is summarized in Table 7.1.

Table 7.1 CMA algorithm summary

Item	Value
Constant term	$R_2 = \frac{E[|x(n)|^4]}{E[|x(n)|^2]}$
Decision signal	$\tilde{x}(n) = \sum\limits_{i=0}^{N} w_i(n)y(n-i)$
Error function	$e(n) = \tilde{x}(n)(R_2 - |\tilde{x}(n)|^2)$
Tap updating algorithm	$\mathbf{W}(n+1) = \mathbf{W}(n) + \mu e(n)\mathbf{Y}^*(n)$
Initial equalizer coefficients	$w(0) = 1,\ i = n;$
	$w(0) = 0,\ i \neq n;$

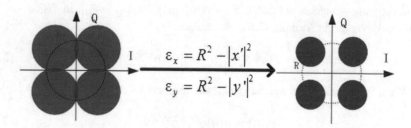

Fig. 7.2 Constellation diagram before and after the equalization of CMA algorithm

We can also see from the above derivation that, for CMA algorithm, the update process of the tap coefficient is only concerned with the statistics of the received signal and the transmitted signal and has nothing to do with the error. Therefore, CMA algorithm does not need a known training sequence in the process of iteration. Figure 7.2 shows the constellation diagram before and after the equalization of CMA algorithm.

CMA algorithm is especially suitable for constant amplitude modulation formats, such as M-order phase-shift keying, and is often the only equalization algorithm. However, for the modulation formats which are not of constant amplitude such as higher order quadrature amplitude modulation (QAM), the time average error of the algorithm cannot be reduced to zero, so it will introduce additional noise after equalization.

7.1.2 CMMA Algorithm

Aiming at the problem that steady-state convergence error of the CMA algorithm for QAM signal is not zero, Zhou et al. introduced a cascade multimode algorithm (CMMA) to solve the problem of steady-state error. In this algorithm, by means of cascading to introduce multiple reference circles, they finally reached the final error rate is very close to zero. Corresponding updating equation of the filter tap weight coefficient can be obtained by the stochastic gradient algorithm as shown below:

$$
\begin{aligned}
h_{xx}(k) &\rightarrow h_{xx}(k) + \mu \varepsilon_x(i) e_x(i) \widehat{x}(i-k) \\
h_{xy}(k) &\rightarrow h_{xy}(k) + \mu \varepsilon_x(i) e_x(i) \widehat{y}(i-k) \\
h_{yx}(k) &\rightarrow h_{yx}(k) + \mu \varepsilon_y(i) e_y(i) \widehat{x}(i-k) \\
h_{yy}(k) &\rightarrow h_{yy}(k) + \mu \varepsilon_y(i) e_y(i) \widehat{y}(i-k)
\end{aligned}
\tag{7.12}
$$

For 8-QAM, $e_{x,y}(i)$ can be obtained from:

$$
e_{x,y}(i) = \text{sign}(Z_{x,y}(i) - A_1) \cdot \text{sign}(Z_{x,y}(i))
\tag{7.13}
$$

For 16-QAM, $e_{x,y}(i)$ can be obtained from:

$$
\begin{aligned}
e_{x,y}(i) &= \text{sign}(C_{x,y}(i)) \cdot \text{sign}(B_{x,y}(i)) \cdot \text{sign}(Z_{x,y}(i)) \\
B_{x,y}(i) &= \left| Z_{x,y}(i) - A_1 \right| \\
C_{x,y}(i) &= \left| B_{x,y}(i) - A_2 \right|
\end{aligned}
\tag{7.14}
$$

In the aforementioned formula, $\text{sign}(x)$ is the signal function and can be expressed as $x/|x|$, and μ is the convergence parameter.

Under the circumstances of 8-QAM and 16-QAM modulation format, multi-mode algorithm has improved the SNR performance significantly compared to CMA algorithm, but it also reduces the robustness of filter convergence process. This is because the CMMA algorithm relies on the correct judgment of the radius of transmitted signal, since the space between different rings for QAM signals is smaller than the smallest symbol interval, so when there is a lot of noise or serious signal distortion, the judgment of ring radius will have a lot of mistakes. One solution is to use CMA algorithm to reach pre-convergence at the beginning, and after the pre-convergence, the system uses multimode algorithm for processing. Because multimode algorithm is backward compatible with constant modulus

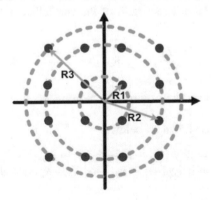

Fig. 7.3 Constellation diagram of 16-QAM signal

algorithm, increasing a CMA process at the beginning will not cause the implementation of complexity. For high-order QAM modulation such as 32-QAM and 64-QAM, the complexity will be very high when multimode algorithm is used for polarization de-multiplexing. We can choose only two or three inner rings for error feedback calculation to reduce complexity. Because the difference of radius between the inner rings of QAM signal is usually larger than the difference of radius of the outer rings, it can also increase the robustness of convergence (Fig. 7.3).

7.1.3 M-CMMA Algorithm

Based on CMA algorithm, researchers proposed the improved CMMA (cascade multimode algorithm). Using cascade approach to introduce multiple reference circle to the equalization algorithm has reduced the signal BER (bit error rate). However, in this algorithm the reference model value of convergence is given by statistical characteristics of signal and does not contain the phase information, so the output of the equalizer is not sensitive to the phase of signal, which will cause the whole phase shift of the output signal; therefore, the output constellation points rotate, so additional phase correction algorithm is eagerly needed. Therefore, in order to recover high-order modulated signal, researchers then put forward the modified cascade multimode algorithm (M-CMMA) using the distribution of signal constellation points. The error function of the equalizer is no longer dependent on a single radius of convergence, but is divided into the in-phase and the orthogonal components, so it has good effect on the recovering of high-order signal. The difference between CMMA and MCMMA is that both the errors of the real and imaginary components are calculated. Thus, the coefficients of the transfer function for I/Q components will be updated individually. On the other hand, the separated cost functions do not take into account of the phase of each symbol, so the subsequent phase rotation is unnecessary. Another benefit from M-CMMA is the reduced number of required reference moduli. And the number of required reference moduli is reduced significantly with the increase in the coding level.

In M-CMMA algorithm, the selected reference radii are $R_1, R_2, R_3 \ldots R_n$; these radii are obtained by the different radii of the constellation diagram of high-order modulation signal:

$$R_1 = \frac{L_1 + L_2}{2}, \; R_2 = \frac{L_3 - L_1}{2}, \; \cdots, \; R_{n-1} = \frac{L_n - L_{n-2}}{2}, \; R_n = \frac{L_n - L_{n-1}}{2} \quad (7.15)$$

where $L_1, L_2, \ldots, L_n$ mean the orthogonal coordinate value of the encoded signal constellation diagram.

The error function is shown as below:

$$\begin{aligned} e_I &= \big|\big|\big|\big|y_I(n)\big| - R_1\big| - R_2\big| \ldots - R_{n-1}\big| - R_n\big| \\ e_Q &= \big|\big|\big|\big|y_Q(n)\big| - R_1\big| - R_2\big| \ldots - R_{n-1}\big| - R_n\big| \end{aligned} \quad (7.16)$$

where $y_I = \text{real}(y(n)), y_Q = \text{imag}(y(n))$.

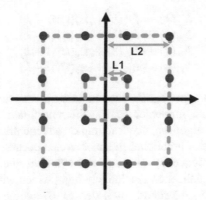

Fig. 7.4 MCMMA constellation diagram

The output signal obtained by M-CMMA algorithm is shown as below:

$$\begin{aligned}
\mathbf{Y}_I(n) &= \mathbf{W}_{11}(n) * \mathbf{X}_I(n) + \mathbf{W}_{12}(n) * \mathbf{X}_Q(n) \\
\mathbf{Y}_Q(n) &= \mathbf{W}_{21}(n) * \mathbf{X}_I(n) + \mathbf{W}_{22}(n) * \mathbf{X}_Q(n)
\end{aligned} \tag{7.17}$$

The updating formula of equalizer tap coefficients is as follows:

$$\begin{aligned}
\mathbf{W}_{11}(n+1) &= \mathbf{W}_{11}(n) + \mu e_I M_I(n) \mathbf{X}_I^*(n) \\
\mathbf{W}_{12}(n+1) &= \mathbf{W}_{12}(n) + \mu e_I M_I(n) \mathbf{X}_Q^*(n) \\
\mathbf{W}_{21}(n+1) &= \mathbf{W}_{21}(n) + \mu e_Q M_Q(n) \mathbf{X}_I^*(n) \\
\mathbf{W}_{22}(n+1) &= \mathbf{W}_{22}(n) + \mu e_Q M_Q(n) \mathbf{X}_Q^*(n)
\end{aligned} \tag{7.18}$$

where

$$M_I(n) = \text{sign}(|\ldots|y_I(i) - R_1| - R_2 \ldots| - R_{n-1}) \ldots \text{sign}(|y_I(i)| - R_1)\text{sign}(y_I(i))$$

$$M_Q(n) = \text{sign}(|\ldots|y_Q(i) - R_1| - R_2 \ldots| - R_{n-1}) \ldots \text{sign}(|y_Q(i)| - R_1)\text{sign}(y_Q(i))$$

$$\tag{7.19}$$

The algorithm diagram is shown below: (Fig. 7.4).

7.1.4   DD-LMS Algorithm

Another way to improve the performance of equalizer SNR is using CMA algorithm for pre-convergence at the beginning, and after that, we use decision-directed least mean square (DD-LMS) algorithm for error calculation. The error computation formula for DD-LMS algorithm is as follows:

$$\varepsilon_{x,y}(i) = Z_{x,y}(i) - d_{x,y}(i) \tag{7.20}$$

where $d_{x,y}(i)$ is the final signal decided by optimum QAM decision border after the recovery of carrier frequency and carrier phase. The updating of filter tap coefficient is based on the following formula:

$$
\begin{aligned}
h_{xx}(k) &\to h_{xx}(k) + \mu \varepsilon_x(i)\hat{x}(i-k) \\
h_{xy}(k) &\to h_{xy}(k) + \mu \varepsilon_x(i)\hat{y}(i-k) \\
h_{yx}(k) &\to h_{yx}(k) + \mu \varepsilon_y(i)\hat{x}(i-k) \\
h_{yy}(k) &\to h_{yy}(k) + \mu \varepsilon_y(i)\hat{y}(i-k)
\end{aligned}
\tag{7.21}
$$

Different from that the pre-equalization and carrier recovery independently used CMA/CMMA algorithm implemented by different functional modules, respectively, CMA/DD-LMS algorithm needs to implement equalization, carrier recovery, and symbol decision in a functional modules/cycle. Because we need to estimate the initial phase and frequency offset of the symbols, the residual phase noise caused by CMA pre-equalization is relatively high, which will to a large extent fail the standard DD-LMS algorithm. In order to overcome this problem, some improvements were made to the DD-LMS algorithm. The improved algorithm uses the error signal which has nothing to do with the phase:

$$
\varepsilon_{x,y}(i) = \left| Z_{x,y}(i) \right|^2 - \left| d_{x,y}(i) \right|^2
\tag{7.22}
$$

From (7.22), we can see the error signal calculation is only based on radius. The improved DD-LMS algorithm is to decide the radius based on optimal QAM decision boundary after carrier frequency and phase recovery. Also, the gap between the QAM rings is smaller than the minimum symbol interval, so the DD-LMS algorithm can achieve better SNR performance than CMMA algorithm. Studies have shown that the difference of 8-QAM and 16-QAM performance is relatively small; however, the performance gap increases with the increase in modulation order.

7.1.5　S-MCMMA Algorithm

Since the M-CMMA algorithm has great effect signals of multi amplitude constellation points, for example, QAM signals, there is still need for the equalization of one-dimensional constellation points such as PAM signals. Based on the M-CMMA algorithm, Zhang for the first time proposed a novel blind equalization algorithm called scalar M-CMMA (S-MCMMA) utilized in the PAM VLC system and experimentally demonstrated a high-speed red-LED-based PAM VLC system employing hardware pre-equalization technology and S-MCMMA equalization.

Since PAM modulation only has real signal conversion and has no imaginary part, the signal error function is one-dimensional. Therefore, we need to make some modification and simplification to M-CMMA and propose a new algorithm suitable for PAM systems, called scalar modified cascaded multimodulus algorithm (S-MCMMA), as shown in Fig. 7.5.

The S-MCMMA error function is presented as follows:

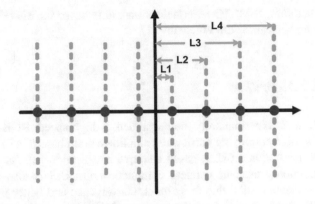

Fig. 7.5 Diagram of S-MCMMA of PAM8 signal

$$\varepsilon = \left|\,\|\,|y(i) - A_1| - A_2|\cdots - A_{n-1}| - A_n\right| \tag{7.23}$$

$$A_1 = \frac{L_1 + L_2}{2},\ A_2 = \frac{L_3 - L_1}{2},\ \cdots,\ A_{n-1} = \frac{L_n - L_{n-2}}{2},\ A_n = \frac{L_n - L_{n-1}}{2} \tag{7.24}$$

where $L_1, L_2, \ldots, L_n$ mean the orthogonal coordinate value of the encoded signal constellation diagram.

Equalizer coefficient and output of the scheme are updated and amended as follows:

$$y(i) = \vec{H}(i) \otimes \vec{X}(i) \tag{7.25}$$

$$M(i) = \text{sign}(|\cdots|y(i) - A_1| - A_2\cdots| - A_{n-1})\cdots\text{sign}(|y(i)| - A_1)\text{sign}(y(i)) \tag{7.26}$$

$$\vec{H}(i+1) = \vec{H}(i) + \mu\varepsilon M(i)\vec{X}^*(i) \tag{7.27}$$

wherein $\vec{X}(i) = [x_{i-N+1}, \ldots x_{i-1}, x_i]$ is the receive signal, namely the input of equalizer; N is the order of equalizer; $\vec{H}(i)$ is each order coefficient for the equalizer; y_i is the current output of equalizer; μ is iterative steps, sign is the sign function, and $\otimes$ is convolution symbol.

Comparing the performance of S-MCMMA equalization and LMS equalization, it can be observed that they behave quite similar (with a slight of advantage of S-MCMMA, especially for PAM 8). However, LMS equalization needs training symbols (about 10% of all symbols) to update and converge the filter weights, while S-MCMMA does not require training symbols in the process of equalization, so it has higher effective information rate compared to LMS. Moreover, after LMS equalization, the constellation point suffers offset and deformable within a certain extent, while S-MCMMA equalization immunes to this problem because it can modify the distorted signal only based on statistical properties of the received signal sequence. Therefore, S-MCMMA equalization can increase the effective data rate and achieve higher bandwidth efficiency [1].

7.1.6 RLS Algorithm

Wang et al. first experimentally demonstrated a high-speed RGB-LED-based WDM VLC system employing carrier-less amplitude and phase (CAP) modulation and recursive least square (RLS)-based adaptive equalization [2]. As an adaptive algorithm commonly used in wireless communication, RLS is also suitable for visible light communication due to its quick convergence and better performance. In [3] and [4], Bandara et al. have investigated RLS-based VLC system and presented the reduction of training sequence for VLC system by employing RLS adaptive equalization. However, the studies are only based on theoretical analysis and numerical simulations, and lack of experimental demonstration.

In high-speed VLC systems, the ISI induced by optical multipath dispersion, sampling time offset, etc., will seriously degrade the system performance and reduce the transmission distance. Considering that the distortion induced by ISI is linear, an equalizer is needed to mitigate the interference and recover the signals. Since the VLC channel is unknown to the receiver, the adaptive algorithm should be used to adapt the filter weights of the equalizer. Owing to its quick convergence and better performance, RLS adaptive algorithm is suitable for VLC systems. The schematic diagram of the RLS-based equalizer is shown in Fig. 7.6. The detailed principle of RLS algorithm has been well described in [3] and [4]. The RLS algorithm recursively finds the filter coefficients that minimize a weighted linear least square cost function related to the input signals. The advantage of the RLS algorithm is that it has quicker convergence than other methods so that the number of training sequence can be greatly reduced. And due to the lower error value, the performance of RLS is better.

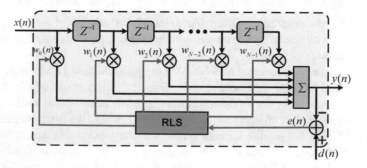

Fig. 7.6 Schematic diagram of RLS-based equalizer

Table 7.2 Required computation of M-CMMA and RLS in one iteration

Algorithm	M-CMMA	RLS
Multiplier	$8N + 16$	$4N^2 + 4N + 1$
Adder	$8N + 20$	$3N^2 + N$
Comparator	28	0
Required iteration	All the symbols	About 2000

In RLS algorithm, $d(n)$ is the desired output of the equalizer. So training sequence is needed to calculate the error value and update the weight vectors. In order to make a computation comparison between RLS and M-CMMA, we calculate the required computation in one iteration as shown in Table 7.2. Here, N is the tap number of the equalizer. It can be found that the required multiplier and adder of M-CMMA are obviously less than RLS in one iteration. The required iteration number of RLS until desired convergence is about 200, while the iteration number of M-CMMA is more than 2000. RLS has better equalization performance and faster convergence speed than M-CMMA at the cost of required computation in one iteration.

Moreover, M-CMMA is a blind equalization algorithm, so the iteration has to be conducted over all the symbols. For RLS, a short training sequence is needed to update and converge the weight vectors until the desired convergence. After that, the iteration is not needed and the symbols are equalized by multiplying the calculated weight vector. Therefore, the whole computation complexity of RLS is much lower than M-CMMA.

On the basis of RLS equalization, Wang et al. an aggregate data rate of 4.5 Gb/s is successfully achieved over 1.5-m indoor free-space transmission with the bit error rate (BER) below the 7% forward error correction (FEC) limit of 3.8×10^{-3}. Moreover, we make a comparison between RLS and MCMMA equalization scheme, and RLS can outperform M-CMMA by Q factor of 1 dB.

7.2 Frequency Domain Equalization Algorithm

In the previous sections, several main time domain equalization algorithms have been introduced, and the corresponding method is the frequency domain equalization, which is from the correction of system frequency characteristic, so that the total characteristic of the baseband system including equalizer satisfies the nondistortion transmission conditions. Compared with time domain equalization, the frequency domain equalization has lower complexity and can reduce the system cost, so it is suitable for the system where the channel is relatively stable. At present, there are two main types of commonly used frequency domain equalization—one is pilot-aided channel estimation, aiming at the OFDM signal, and the other one is the single-carrier frequency domain equalization (SC-FDE) aiming at single-carrier signal. In this section, the two methods will be introduced in details.

7.2.1   Pilot-Aided Channel Estimation Algorithm

We usually think that in OFDM systems, the channel response within a symbol remains the same. When considering system design, it is guaranteed that length of the cyclic prefix is greater than the maximum path delay of the channel. When the cyclic prefix length is greater than the maximum channel delay, the essence of the frequency domain equalizer is by adding and removing the cyclic prefix to transform the channel response that the transmitted signal undergoes into a cyclic matrix. At the receiver end, the frequency domain signals after Fourier transform strictly equal to the product of the transmitted frequency domain signals and frequency domain channel response, so to compensate the channel directly in the frequency domain can recover the transmitted frequency domain signals. Channel estimation based on pilot means inserting a certain percentage of pilot frequency into data, it has simple structure, and the algorithm complexity is small, but the pilot information needs to be inserted. For pilot-symbol-aided channel estimation, the issue to consider is how to choose the suitable means of pilot insertion and how to use the least amount of pilot to get the most accurate channel response. Three typical pilot insertion schemes are block pilot, comb pilot, and discrete pilot, respectively, as shown in Fig. 7.7.

Wherein block pilot is suitable for the slow fading channel, the comb pilot is suitable for the rapid changing channel, and discrete pilot has the advantages of the above two pilot and can maximum saving pilot overhead. There are two commonly used methods for pilot position channel estimation, including channel estimation based on LS and channel estimation based on the MMSE.

The response of pilot location channel estimation based on the LS is as follows:

$$\hat{H}_p = X_p^{-1} Y_P \tag{7.28}$$

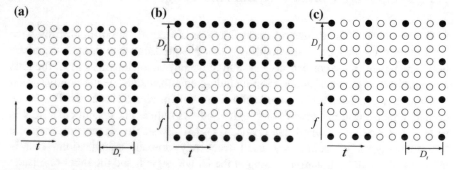

Fig. 7.7 Typical pilot insertion scheme **a** block pilot; **b** comb pilot; **c** discrete pilot

It can be seen from the formula that we need a division operation in the position of pilot subcarrier, and then we can get the frequency domain channel response of the location of the pilot. However, the LS algorithm does not employ the correlation of channel in the time domain and frequency domain and ignore the influence of noise N to H_p. In fact, the channel estimation is relatively sensitive to the noise. When SNR is low, the accuracy of LS estimate will be greatly reduced, thus affecting the performance of the system.

Based on MMSE channel estimation, the response of the pilot location is as follows:

$$\begin{aligned}
\hat{H}_p &= R_{HY}R_{YY}^{-1}Y_P \\
&= R_{HH}[R_{HH} + \sigma_n^2(X_p^H X_p)]^{-1}X_P^{-1}Y_P \\
&= R_{HH}[R_{HH} + \sigma_n^2(X_p^H X_p)]^{-1}\hat{H}_{P,LS}^{-1}
\end{aligned} \tag{7.29}$$

In order to estimate the channel response of the pilot position, we should not only know the correlation value between the received signals on the pilot position, but also need to know the second-order statistical properties of the channel. In actual system, we may need to presuppose a most likely channel model to calculate the autocorrelation R_{HH} of channel response matrix. The structure of MMSE algorithm is complex, when the channel model is matched, performance in statistical sense is optimal, but once the model mismatches, there will be a larger estimation error. After calculating the channel response of pilot location, interpolation method is used to estimate of channel response of all the position. Finally, according to the channel response matrix algorithm use forcing zero algorithm, MMSE algorithm or iterative equalization algorithm to find the channel compensation matrix **W**.

7.2.2 SC-FED Algorithm

The single-carrier modulation technology based on frequency domain equalization (SC-FDE) is a kind of modulation technology with high spectrum efficiency based on single carrier [5]. High PAPR is a fatal weakness for serious nonlinear visible optical communication system, so the SC-FDE has certain advantages compared to OFDM; the spectrum efficiency is the same with OFDM modulation technology, and the complexity of them is consistent, but SC-FDE has a smaller PAPR [6]. SC-FDE technique and OFDM modulation are generally consistent, and the difference is that the FFT operation is moved from the transmitter to the receiver (Fig. 7.8).

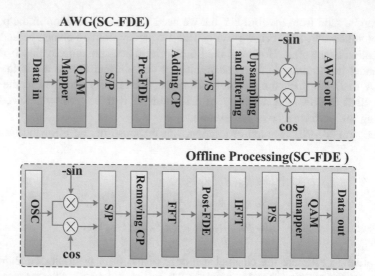

Fig. 7.8 Block diagram of modulation/demodulation principle of SC-FDE technology. S/P: serial–parallel conversion; Pre-FDE: pre-equalization in frequency domain; Adding CP: adding cyclic prefix; P/S: parallel–serial conversion; Removing CP: removing cyclic prefix; FFT: fast Fourier transform; Post-FDE: post-equalization in frequency domain

7.3   Nonlinear Equalization Algorithm

7.3.1   Volterra Series Algorithm

In high-speed VLC system, the LED nonlinearity is mainly induced by the non-linear relationship between the LED forward current and the bias voltage, which will seriously degrade the system performance and reduce the transmission distance. An equalizer is needed to mitigate the interference and recover the signals. Volterra series expansion is a widely used nonlinear system representation to model most nonlinear systems. Therefore, Volterra series-based nonlinear equalizer is suitable to mitigate the LED nonlinearity. The schematic diagram of the Volterra series-based equalizer is shown in Fig. 7.9.

The Volterra series expansion contains a linear term and nonlinear series. The linear term is utilized for linear equalizer, while the nonlinear series consist of second-order term, third-order term, etc. As a trade-off between computation complexity and equalization performance, only the second-order term is taken into consideration in our VLC system, and the higher order terms are omitted. So the output of the equalizer is expressed as:

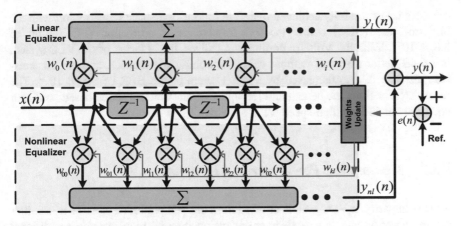

Fig. 7.9 Schematic diagram of Volterra series-based equalizer

$$y(n) = y_l(n) + y_{nl}(n)$$
$$= \underbrace{\sum_{i=0}^{N-1} w_i(n)x(n-i)}_{y_l(n)} + \underbrace{\sum_{k=0}^{NL-1}\sum_{l=k}^{NL-1} w_{kl}(n)x(n-k)x(n-l)}_{y_{nl}(n)} \qquad (7.30)$$

Here $y_l(n)$ is the output of the linear equalizer, and $y_{nl}(n)$ is the output of the nonlinear equalizer. N and NL are the tap numbers of the linear and the nonlinear equalizer. $w_i(n)$ and $w_{kl}(n)$ are the weights of the linear and nonlinear equalizers. In our VLC system, the optimal tap numbers of the linear and the nonlinear equalizers are, respectively, 45 and 25.

Yiguang Wang used M-CMMA to calculate the error function and update the weights of the nonlinear equalizer. M-CMMA is a blind multimodulus algorithm, so training symbol is not required for convergence. The weights of the nonlinear equalizer are updated as follows:

$$w_{kl\_11}(n+1) = w_{kl\_11}(n) + \mu\varepsilon_I M_I \cdot x_I(n-k)x_I(n-l)$$
$$w_{kl\_12}(n+1) = w_{kl\_12}(n) + \mu\varepsilon_I M_I \cdot x_Q(n-k)x_Q(n-l)$$
$$w_{kl\_21}(n+1) = w_{kl\_21}(n) + \mu\varepsilon_Q M_Q \cdot x_I(n-k)x_I(n-l) \qquad (7.31)$$
$$w_{kl\_22}(n+1) = w_{kl\_22}(n) + \mu\varepsilon_Q M_Q \cdot x_Q(n-k)x_Q(n-l)$$

They experimentally demonstrate a high-speed WDM CAP64 VLC system employing Volterra series-based nonlinear equalizer to mitigate the LED nonlinearity. M-CMMA is proposed to calculate the error function and update the weights of the nonlinear equalizer without using training symbols. An aggregate data rate of

4.5 Gb/s is successfully achieved over 2-m indoor free-space transmission with the bit error rate (BER) below the 7% forward error correction (FEC) limit of 3.8×10^{-3}. With the Volterra nonlinear equalizer, the Q factor of the VLC system is 1.6 dB better than that without using the nonlinear equalizer, and the transmission distance is also increased by about 110 cm at the BER of 3.8×10^{-3}. The results clearly show the benefit and feasibility of the Volterra series-based nonlinear equalizer for indoor high-speed VLC systems [7].

7.3.2   Memoryless Power Series Algorithm

The nonlinearity in VLC system aroused from the PIN photodetector, the transmitter driving circuits, and the receiver amplification circuits may introduce additional nonlinear noises and can have detrimental effects to the signal reception. Therefore, it is an essential issue to measure the statistic characteristics of the nonlinear response of the whole VLC system to achieve a better fitting compensation. As a trade-off between computational complexity and pre-distortion performance, channel memory depth is not considered in our VLC system. The adaptability of MPS-based pre-distorter is implemented based on training symbols. Assuming the signal after IFFT in the transmitter signal processing is $x(n)$, n represents time index and the receiving signal without pre-distortion is $y(n)$. K represents channel nonlinearity order. To implement adaptive MPS-based pre-distortion, channel polynomial coefficients estimation and inversion are required. Using training symbols and formula (7.29), the coefficients D_k of MPS-based pre-distorter are estimated to solve the problem.

$$x(n) = D_l y_s(n)|y_s(n)| + \cdots + D_{k-l} y_s(n)|y_s(n)|^{k-l} = \sum_{k=0}^{K-l} D_k y_s(n)|y_s(n)|^k \quad (7.32)$$

In formula (7.29), $y_s(n)$ represents the normalized and time-shifted received signal without pre-distortion. Thus, $x(n)$ is pre-distorted to be $x_d(n)$ and the receiving signal after pre-distortion is $y_d(n)$ which are shown in Fig. 7.10.

A novel MPS-based adaptive nonlinear pre-distortion scheme to mitigate nonlinear impairments is proposed for high-speed VLC system. The nonlinear compensation function is merely related to current sampling point; therefore, the computation complexity is greatly reduced. System performance improvement employing MPS-based adaptive nonlinear pre-distorter is confirmed through experimental demonstration. They have successfully demonstrated 1.6 Gbit/s 16-QAM-OFDM VLC system utilizing MPS-based pre-distorter [8].

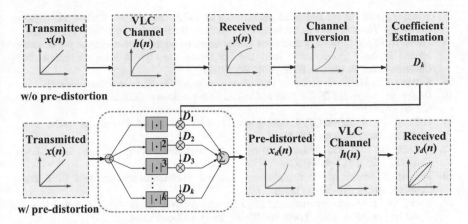

Fig. 7.10 Principle of MPS-based adaptive nonlinear pre-distorter

7.4 Summary

This chapter is around the post-equalization technology in visible light communication system and mainly introduced the time domain equalization algorithm and frequency domain equalization algorithm. Equalization technique plays an important role for high-speed visible light communication system; it can promote the maximum transmission rate and reliability of the system and is the important guarantee to realize high-speed and long-distance visible light transmission.

References

1. Zhang, M., Wang, Y., Wang, Z., et al.: A novel scalar MCMMA blind equalization utilized in 8-PAM LED based visible light communication system. In: IEEE International Conference on Communications Workshops. IEEE, pp. 321–325 (2016)
2. Wang, Y., Huang, X., Tao, L., et al.: 4.5-Gb/s RGB-LED based WDM visible light communication system employing CAP modulation and RLS based adaptive equalization. Opt. Express **23**(10), 13626–13633 (2015)
3. Bandara, K., Niroopan, P., Chung, Y.: Improved indoor visible light communication with PAM and RLS decision feedback equalizer. Inst. Electron. Telecommun. **59**(6), 672–678 (2013)
4. Bandara, K., Chung, Y.: Reduced training sequence using RLS adaptive algorithm with decision feedback equalizer in indoor visible light wireless communication channel. In: IEEE International Conference on ICT Convergence (ICTC, 2012), pp. 149–154
5. Wang, Y., Shi, J., Yang, C., Wang, Y., Chi, N.: Integrated 10Gb/s multi-level multi-band PON and 500 Mb/s indoor VLC system based on N-SC-FDE modulation. Opt. Lett. **39**(9), 2576–2579 (2014)
6. Wang, Y., Huang, X., Zhang, J., Wang, Y., Chi, N.: Enhanced performance of visible light communication employing 512-QAM N-SC-FDE and DD-LMS. Opt. Exp. **22**(13), 15328–15334 (2014)

7. Wang, Yiguang, et al.: Enhanced performance of a high-speed WDM CAP64 VLC system employing Volterra series-based nonlinear equalizer. IEEE Photonics J. **7**(3), 1–7 (2015)
8. Zhou, Y., et al.: A novel memoryless power series based adaptive nonlinear pre-distortion scheme in high speed visible light communication. In: Optical Fiber Communications Conference and Exhibition IEEE, W2A.40 (2017)
9. Kaminow, I., Li, T.: Optical fiber telecommunications IVA. Elsevier Science (2002)
10. Oerder, M., Meyr, H.: Digital filter and square timing recovery. IEEE Trans. Commun. **36**(5), 605–612 (1988)
11. Gardner, F.: A BPSK/QPSK timing-error detector for sampled receivers. IEEE Trans. Commun. **34**(5), 423–429 (1986)
12. Godard, D.: Passband timing recovery in an all-digital modem receiver. IEEE Trans. Commun. **26**(5), 517–429 (1978)
13. Mueller, K., Muller, M.: Timing recovery in digital synchronous data receivers. IEEE Trans. Commun. **24**(5), 516–531S (1976)
14. Viterbi, A.: Nonlinear estimation of PSK-modulated carrier phase with application to burst digital transmission. IEEE Trans. Inf. Theor. **29**(4), 543–551 (1983)
15. Peveling, R., Pfau, T., Aamczyk, O., Eickhoff, R., Noe, R.: Multiplier-free real-time phase tracking for coherent QPSK receivers. IEEE Photonics Technol. Lett. **21**(3), 137–139 (2009)
16. Pfau, T., Hoffmann, S., Noe, R.: Hardware-efficient coherent digital receiver concept with feedforward carrier recovery for -QAM constellations. J. Lightwave Technol. **27**(8), 989–999 (2009)
17. Zhou, X.: An improved feed-forward carrier recovery algorithm for coherent receivers with M-QAM modulation format. IEEE Photon. Technol. Lett. **22**(14), 1051–1053 (2010)
18. Gao, Y., Lau, A., Lu, C., Wu, J., Li, Y., Xu, K.: Low-complexity two-stage carrier phase estimation for 16-QAM systems using QPSK partitioning and maximum likelihood detection. In: Optical Fiber Communication Conference and Exposition (OFC/NFOEC), 2011 and the National Fiber Optic Engineers Conference, pp. 1–3 (2011)
19. Bülow, H., et al.: Measurement of the maximum speed of PMD fluctuation in installed field fiber. In: Optical Fiber Communication Conference, 1999, and the International Conference on Integrated Optics and Optical Fiber Communication. OFC/IOOC'99. Technical Digest, vol. 2, pp. 83–85 (1999)
20. Louchet, H., Kuzmin, K., Richter, A.: Improved DSP algorithms for coherent 16-QAM transmission. In: 34th European Conference on Optical Communication, 2008. ECOC 2008, pp. 1–2 (2008)
21. Zhou, X., Yu, J., Magill, P.: Cascaded two-modulus algorithm for blind polarization demultiplexing of 114-Gb/s PDM-8-QAM optical signals. In: Optical Fiber Communication Conference. Optical Society of America (2009)
22. Spalvieri, A., Valtolina, R.: Data-aided and phase-independent adaptive equalization for data transmission systems. European Patent Application EP 1.089: 457
23. Wang, Y., Huang, X., Zhang, J., Wang, Y., Chi, N.: Enhanced performance of visible light communication employing 512-QAM N-SC-FDE and DD-LMS. Opt. Express **22**(13), 15328–15334 (2014)
24. Wang, Y., Shi, J., Yang, C., Wang, Y., Chi, N.: Integrated 10Gb/s multi-level multi-band PON and 500Mb/s indoor VLC system based on N-SC-FDE modulation. Opt. Express **39**(9), 2576–2579 (2014)

Chapter 8
High-Speed VLC Communication System Experiments

Speed is an important indicator of a communication system, and providing high-speed wireless access is also a significant advantage of visible light communication. At present, the transmission rate of Wi-fi is about 100 Mbit/s, and the power line communication is basically at this rate and even lower. Compared with traditional wireless access technology, visible light communication can realize more high-speed wireless access due to its potentially huge bandwidth. At present, international research hotspots are focusing on high-speed transmission of visible light using high-order modulation, multiplexing, and networking. This chapter will focus on the high-speed VLC system, from advanced modulation technology, two-way transmission and multi-user access, multidimensional multiplexing, MIMO, and visible light network, to introduce the corresponding technical principle and high-speed VLC experimental results.

8.1 Advanced Modulation Technology in VLC System

In visible light communication system, LED modulation bandwidth is very limited. Currently, the 3 dB bandwidth of commercial LED is only a few megahertz. In order to improve the transmission rate of the system, it is important to adopt advanced modulation technology with high spectral efficiency in addition to extending its bandwidth from LED structure and driving circuit design. This section will focus on the principles, implementation, and advantages of OFDM, CAP, and SC-FDE and introduce the high-speed VLC experimental system with corresponding modulation technology.

8.1.1 Single-Carrier Modulation Based on Frequency Domain Equalization

Single-carrier modulation based on frequency domain equalization (SC-FDE) is a high-spectrum efficiency modulation based on single carrier. High PAPR is a fatal flaw in the nonlinear severe visible light communication system. Therefore, SC-FDE has certain advantages over OFDM, which has same spectral efficiency and same complexity with OFDM, but has a lower PAPR. The modulation/demodulation schematic diagram of SC-FDE is shown in Fig. 8.1.

As seen from the SC-FDE diagram, the modulation technique is basically consistent with OFDM, except that IFFT is moved from the transmitter to the receiver.

Based on the above SC-FDE modulation technology, combined with advanced equalization algorithm, we achieve the experiment of 4.22 Gb/s in VLC transmission system. The experimental setup is shown in Fig. 8.2. In experiment, we adopt single-carrier modulation technology, select RGB-LED as light source, and finally achieve WDM visible light transmission. At the transmitter, pre-equalization is used to compensate the frequency fading of the LED. At the same time, the time–frequency domain hybrid equalization technology combined with FDE and DD-LMS is utilized at receiver. This is the highest rate of visible light communication transmission currently reported internationally.

In order to improve the spectrum efficiency, 512-QAM high-order modulation is adopted in the experiment. The modulation bandwidth of each wavelength is 156.25 MHz. The measured spectrum is shown in Fig. 8.3.

The value of measured Q factor (dB) versus modulation order is presented in Fig. 8.4. It can be seen that the Q factor performance of three wavelength signals of red, green, and blue has been improved by 1.4, 1.6, and 1 dB, respectively, after

Fig. 8.1 Modulation/
demodulation schematic
diagram of SC-FDE

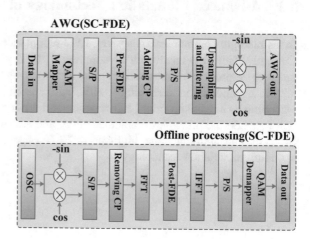

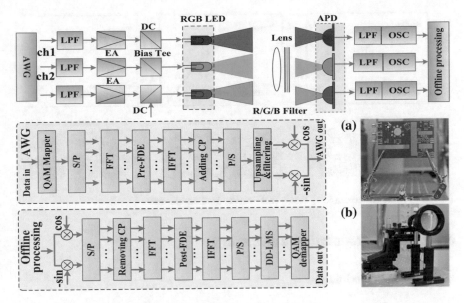

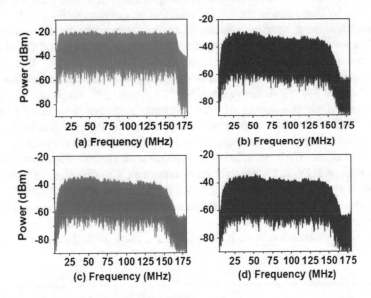

Fig. 8.2 Experimental setup of SC-FDE VLC system

Fig. 8.3 Measured spectrum of 512-QAM WDM

using time–frequency mixed equalization technique. The experimental results verify that the SC-FDE modulation technology has the advantages of simple structure, low computational complexity, and high spectral efficiency, which shows a wide application prospect in high-speed visible light communication system.

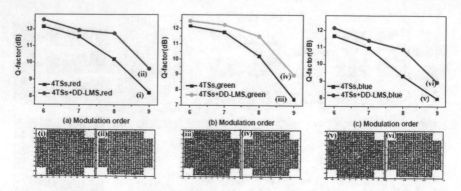

Fig. 8.4 Value of measured Q factor (dB) versus modulation order

8.1.2   CAP Modulation Technology

8.1.2.1   CAP Modulation

CAP (carrier-less amplitude and phase, no carrier and amplitude) modulation method is a kind of multidimensional multistage modulation technology; it is firstly put forward by Bell Labs in the 1970s. With this modulation technique, high-speed transmission of high spectral efficiency can be achieved under the condition of limited bandwidth. Compared with traditional QAM modulation and OFDM, CAP modulation adopts two orthogonal digital filters. The advantage is that CAP no longer requires the plural of electrical or optical signal modulation to the real signal conversion, and this conversion usually needs a mixer, radio frequency source, or a light IQ modulator to achieve. At the same time, compared with OFDM modulation, the CAP modulation no longer needs to adopt the discrete Fourier transform (DFT) as well, thus greatly reducing the computational structural complexity of the system. Therefore, CAP modulation is applicable to systems requiring low complexity, such as PON, in visible light communication. The structures of typical CAP modulation system transmitter and receiver are shown in Figs. 8.5 and 8.6.

Seen from the figure, CAP modulation adopts two mutually orthogonal filters in transmitter, by controlling the shaping filter coefficient of signal and the order number. CAP modulation with high-order has a narrow bandwidth and does not

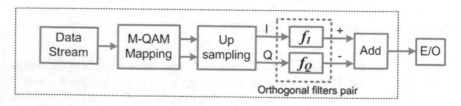

Fig. 8.5 Structure of transmitter in CAP modulation system

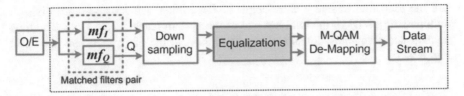

Fig. 8.6 Structure of receiver in CAP modulation system

require mixer. At the receiver, CAP modulation restores signal by adaptive filter. As a result, CAP modulation has a simple structure as its advantage.

The CAP modulation signal can be expressed as follows:

$$s(t) = a(t) \otimes f_1(t) - b(t) \otimes f_2(t) \tag{8.1}$$

Here, $a(t)$ and $b(t)$ are signals from the original bits of the I and Q paths that are encoded and sampled. $f_1(t) = g(t)\cos(2\pi f_c t)$ and $f_2(t) = g(t)\sin(2\pi f_c t)$ are the time domain function of the corresponding formed filter, which forms a pair of Hilbert transform pairs.

Assuming the transmission channel is ideal, the output of two matching filters on the receiver can be expressed as follows:

$$\begin{aligned} r_i(t) &= s(t) \otimes m_1(t) = (a(t) \otimes f_1(t) - b(t) \otimes f_2(t)) \otimes m_1(t) \\ r_q(t) &= s(t) \otimes m_2(t) = (a(t) \otimes f_1(t) - b(t) \otimes f_2(t)) \otimes m_2(t) \end{aligned} \tag{8.2}$$

Here, $m_1(t) = f_1(-t)$ and $m_2(t) - f_2(-t)$ are the corresponding pulse response of the matched filter. The corresponding matching filter can be used to demodulate the original signal at the receiver.

8.1.2.2   Experiment of CAP Modulation

(1)   The experiment of 3-sub CAP

CAP modulation has great application value in visible light communication due to its simple structure and low computational complexity. The visible light communication with CAP modulation technology has been verified by experiment. The experimental structure of visible light communication based on CAP modulation is shown in Fig. 8.7.

In this experiment, RGB-LED is adopted as the light source to realize WDM VLC system. At the same time, the experiment also adopts pre-equalization and post-equalization technology based on M-CMMA to improve the frequency response and system performance of RGB-LEDs. Frequency division multiplexing (FDM) technology is used in every wavelength. We, respectively, modulate signals of different users on three subcarriers. Each subcarrier modulation signal bandwidth is 25 MHz with 64-QAM modulation, so data rate of each subcarrier is 150 Mbit/s, and the total rate of every wavelength is 450 Mbit/s. After wavelength division multiplexing (WDM), the total net rate of the system reaches 1.35 Gb/s. The time domain and frequency domain of the filter used in the experiment are shown in Fig. 8.8.

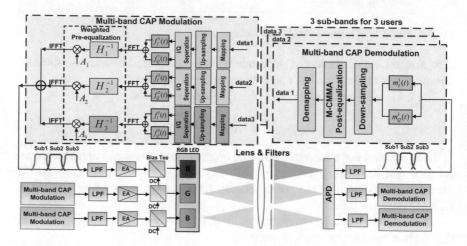

Fig. 8.7 Experimental structure of VLC system based on CAP modulation

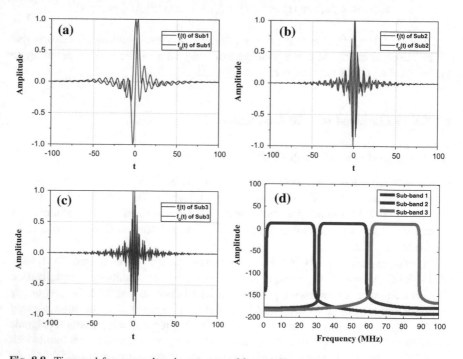

Fig. 8.8 Time and frequency domain response of formed filter

It can be seen from the experimental BER curve that CAP modulation has the characteristic with simple structure, low computational complexity, and high spectral efficiency. It is an important modulation technique to realize high spectral efficiency and high-speed transmission in limited bandwidth resources and has great potential in the visible light communication.

(2) **The experiment of 8-Gb/s RGBY LED-based WDM VLC system employing high-order CAP modulation and hybrid post-equalizer**

In this section, we propose to use a hybrid post-equalizer in a high-order carrier-less amplitude and phase (CAP) modulation-based visible light communication (VLC) system. The hybrid equalizer consists of a linear equalizer, a Volterra series-based nonlinear equalizer, and a decision-directed least mean square (DD-LMS) equalizer in order to simultaneously mitigate the linear and nonlinear distortions of the VLC system. A commercially available RGBY LED is utilized for four wavelengths multiplexing. By the hybrid equalizer, an aggregate data rate of 8 Gb/s is experimentally achieved over 1-m indoor free-space transmission with the bit error rate (BER) below the 7% forward error correction (FEC) limit of 3.8×10^{-3}. To the best of our knowledge, this is the highest data rate ever reported in high-speed VLC systems.

In high-speed VLC systems, there exist severe linear and nonlinear distortions. The linear distortions induced by optical multipath dispersion, sampling time offset, etc., will result in the ISI, while the LED nonlinearity will cause the distortion of signal constellations. To simultaneously mitigate the linear and nonlinear distortions and recover the signals, we propose the use of a hybrid post-equalizer, as shown in Fig. 8.9.

It can be found that the hybrid equalizer consists of two-stage filters. The first stage filter includes a linear equalizer and a Volterra series-based nonlinear equalizer. M-CMMA is used to calculate the error function and update the weights of the linear and nonlinear equalizer [1]. The output of the first stage filter can be expressed as:

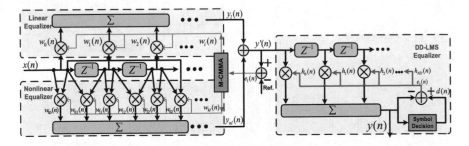

Fig. 8.9 Schematic diagram of the proposed hybrid equalizer

$$y'(n) = y_l(n) + y_{nl}(n)$$

$$= \underbrace{\sum_{i=0}^{N-1} w_i(n)x(n-i)}_{y_l(n)} + \underbrace{\sum_{k=0}^{NL-1} \sum_{l=k}^{NL-1} w_{kl}(n)x(n-k)x(n-l)}_{y_{nl}(n)}. \qquad (8.3)$$

Here, $y_l(n)$ is the output of the linear equalizer and $y_{nl}(n)$ is the output of the nonlinear equalizer. N and NL are the tap numbers of the linear and the nonlinear equalizer. $w_i(n)$ and $w_{kl}(n)$ are the weights of the linear and nonlinear equalizer. In our VLC system, the optimal tap numbers of the linear and the nonlinear equalizer are, respectively, 45 and 25 [2].

Then, the output of the first stage filter is sent into the second stage filter. The second stage filter is a DD-LMS equalizer to obtain good performance after the pre-convergence. DD-LMS is a stochastic gradient descent algorithm and does not depend on the statistics of symbols but relies on the symbol decisions. The detailed principle of DD-LMS has been well described in [3]. Therefore, the output of the DD-LMS equalizer can be written as:

$$y(n) = \sum_{i=0}^{ND-1} h_i(n)y'(n-i). \qquad (8.4)$$

Here, $h_i(n)$ and ND are, respectively, the weights and the tap number of the DD-LMS equalizer. The optimal tap number of the DD-LMS equalizer is 33 in the VLC system [4].

Figure 8.10 shows the experimental setup of the RGBY LED-based WDM VLC system employing high-order CAP modulation and the proposed hybrid post-equalizer. At the transmitter, the original bit sequence is firstly mapped into

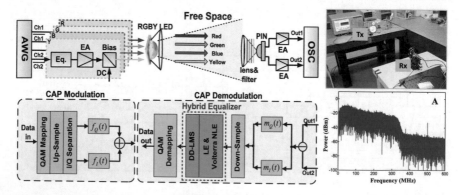

Fig. 8.10 Experimental setup of the WDM VLC system employing high-order CAP and the hybrid post-equalizer

complex symbols. Then, the high-order QAM signal is sent for CAP modulation. The detail of the CAP modulation and demodulation has been well described in [5]. Here, $f_I(t)$ and $f_Q(t)$ are the orthogonal shaping filter pair. The roll-off coefficient of the square-root-raised cosine function for CAP modulation is set to 0.01 for a high spectral efficiency.

In this experiment, we use Tektronix AWG 7122C to generate the CAP signals for the four color chips of the RGBY LED. The AWG 7122C has two independent outputs, so we use the output of channel 1 and its inverted output for the red and green chip, while the output of channel 2 and its inverted output are used for the blue and yellow chip, respectively. The modulation bandwidth of each color chip is fixed at 320 MHz. The generated CAP signals are then pre-amplified by a self-designed bridged-T-based pre-equalizer to compensate the LED frequency attenuation at high-frequency component [6]. The received electrical spectrum after the pre-equalizer is shown in the inset A of Fig. 8.10. Through an electrical amplifier (EA, Minicircuits, 25-dB gain), the electrical signal and DC bias voltage are combined by a bias tee, respectively, and used to drive the four color chips of the RGBY LED. Here, a commercially available RGBY LED (LED Engine, output power: 1 W) is utilized as the transmitter. A reflection cup with 60° divergence angle is applied to the RGB-LED to decrease the beam angle of the LED for longer transmission distance.

At the receiver, a commercial PIN photodiode (Hamamatsu 10784) is used to detect the optical signals. Before the PIN, lens (50 mm diameter and 50 mm focus length) is used to focus light, and optical R/G/B/Y filters are also employed to filter out different colors. Here, we design a differential receiving circuit for the PIN, so that two received differential signals (Out1 and Out2) are obtained. The differential outputs of the receiver are amplified by EAs and then recorded by two different channels of a digital storage oscilloscope (Agilent DSO54855A) for further off-line demodulation and signal processing.

In off-line signal processing, the subtraction is firstly operated between the received differential signals to increase the SNR. Then, the subtracted signal is sent for CAP demodulation. Accordingly, $m_I(t)$ and $m_Q(t)$ are the matching filter pair for demodulation. After demodulation, the proposed hybrid post-equalizer is employed, and QAM decoder is followed to further recover the original bit sequence.

To render the RGBY LED working at the optimal condition, we firstly study the influence of different bias voltages and input signal peak-to-peak voltage (Vpp). The measured BER performance of the red chip versus different bias voltages is shown in Fig. 8.11a. At this time, the input signal Vpp is fixed at 0.4 V. Then, the BER performance versus different input signal Vpps is measured at the fixed bias voltage of 1.85 V, as shown in Fig. 8.11b. It can be found that compared to using only the linear and nonlinear equalizer, the best BER performance can be obtained by the hybrid equalizer. Moreover, the performance improvement is much better at higher bias voltage and larger signal Vpp. It can be explained by noting that the utilized Volterra nonlinear equalizer can bring better performance at higher bias voltage and larger signal Vpp, because of the mitigation of the LED nonlinearity. We also

Fig. 8.11 Measured BER of the red chip versus **a** different bias voltages and **b** different input signal Vpp

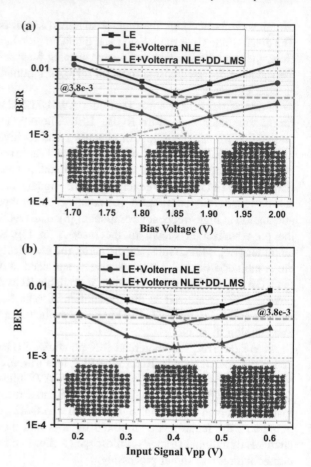

measure the BER performances versus bias voltages and input signal Vpps of the other three color chips. According to the measurements, the optimal working points of the red, green, blue, and yellow chip are, respectively, at (1.85 V bias voltage, 0.4 V input signal Vpp), (3.0, 0.45 V), (2.9, 0.35 V), and (2.1, 0.8 V).

At the optimal working point, we measure the BER performances of the four color chips versus different modulation orders, as shown in Fig. 8.12. The measurement is operated at 1-m transmission distance. It is observed that the highest modulation orders meeting the 7% FEC threshold of 3.8×10^{-3} for the four color chips (RGBY) are 128-QAM, 32-QAM, 64-QAM, and 128-QAM, respectively. Therefore, the aggregate data rate of $320 \times (7 + 5 + 6 + 7) = 8$ Gb/s is successfully achieved at a distance of 1 m. It is worth noting that the red chip has better performance because the best responsivity of the utilized PD is at 620 nm, which is close to the red light wavelength. It means that when the four color channels are modulated by the same order CAP signals, the red chip will have the best BER performance. In other words, the red chip can support the transmission of higher order CAP signal (128-QAM). Therefore, to maximum VLC system capacity, we

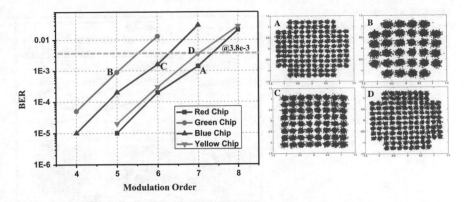

Fig. 8.12 BER performance versus different modulation orders of the four color chips

choose 128-QAM CAP signal for the red chip, while 128-QAM for yellow, 64-QAM for blue, and 32-QAM for green. In this case, the performance of 128-QAM red chip is not the best (little worse than 32-QAM green chip).

The BER performance versus different distances employing the proposed hybrid equalizer is presented in Fig. 8.13. We, measure the BER performances at 0.5, 1, 1.5, and 2 m respectively. It can be observed that at the distance of 0.5 and 1 m, the values of BER of all the four color chips are below the 7% FEC limit of 3.8×10^{-3}, while at the distance of 1.5 m only the yellow chip cannot meet the 7% FEC requirement.

In our experiment, the CAP signals are independently modulated onto the four color channels. Then, the four color chips of the RGBY LED are lighted to simultaneously transmit signals. In order to investigate the cross talk for each channel, we measure the BER performances of the red chips without the other three color chips, as shown in Fig. 8.13 (the black line). Compared to the performance of

Fig. 8.13 BER performance versus different transmission distances

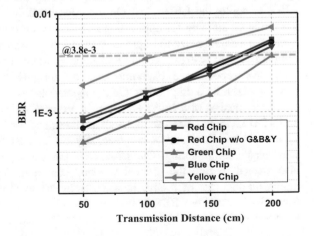

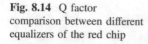

Fig. 8.14 Q factor
comparison between different
equalizers of the red chip

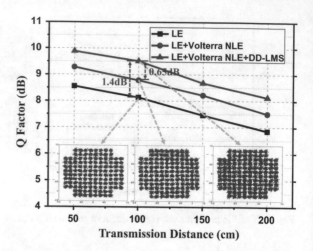

the other three channels (the red line), it can be clearly seen that there is almost no
cross talk induced by other three channels. It is because the wavelengths of the four
color channels are different, and we use the R/G/B/Y filters in front of the PIN.
Therefore, the cross talk from other channels has been filtered out before detection.

In order to make a clear comparison, we measure the Q factor of the red chip
versus different transmission distances with the linear equalizer, the linear equalizer + the Volterra nonlinear equalizer, and the proposed hybrid equalizer,
respectively, as shown in Fig. 8.14. The results show that the hybrid equalizer can
outperform the linear equalizer + the Volterra nonlinear equalizer by 0.65 dB and
outperform the linear equalizer by 1.4 dB at the distance of 1 m.

It should be noted that in VLC system the luminance of the LED is the key factor
that limits the transmission distance. In our experiment, the measured luminance of
the RGBY LED at 1 m is about 450 lx. The illumination level is below the standard
value for brightness (500 lx). So, it is believed that transmission distance and
system performance can be further improved by increasing the optical power of
LEDs or deploying LED array. On the other hand, it is very important for a VLC
system to be suitable for applications. Now, we are trying to design an auto-focus
system in a smaller size to improve its practicality, which integrates the lens and the
optical filters to realize focusing and filtering simultaneously.

In this section, we experimentally demonstrate the feasibility of a hybrid
post-equalizer in a high-order CAP modulation-based VLC system. The hybrid
equalizer consists of a linear equalizer, a Volterra series-based nonlinear equalizer,
and a DD-LMS equalizer, so that both the linear and nonlinear distortions of the VLC
system can be eliminated. A commercially available RGBY LED is utilized for four
wavelengths multiplexing. The influence of the bias voltages and input signal Vpp is
also studied to find the optimal working condition. By the hybrid equalizer, an
aggregate data rate of 8 Gb/s is successfully achieved over 1-m indoor free-space
transmission with the BER below the 7% FEC limit of 3.8×10^{-3}. To our best
knowledge, this is the highest data rate ever obtained in high-speed VLC systems.

8.1.3   *Orthogonal Frequency Division Multiplexing (OFDM)*

8.1.3.1   OFDM Modulation

Orthogonal frequency division multiplexing (OFDM) is a new and efficient coding technique, which is a kind of multicarrier modulation. It can effectively resist multipath interference, which can still reliably receive the signal of interference, and its signal frequency band utilization is also greatly improved. In 1971, Weinstein and Eben put forward using the discrete Fourier transform to achieve modulation and demodulation in OFDM system. It simplified the strict synchronization problem between the oscillator array and the local carrier in relevant receiver. As a result, it laid theoretical foundation for achieving the fully digital scheme of OFDM. After the 1980s, with the development of digital signal processing (DSP) technology and increasing demand for high-speed data communication, the modulation technology of OFDM was once again a hotspot. OFDM technology is getting more and more attention because it has many unique advantages:

(1) High spectral efficiency. The spectral efficiency is nearly twice as high as the serial system. It is important in a wireless environment where spectrum resources are limited.
(2) Strong ability to resist multipath interference and frequency selective fading. Because the OFDM system spread out the data into many subcarriers and greatly reduces the symbol rate of each subcarrier to abate the effect of multipath propagation, it can even eliminate inter-symbol interference caused by multipath by using cyclic prefix as a means of protection interval.
(3) Achieve the maximum bit rate by adopting dynamic subcarrier distribution technology. By selecting each subchannel, the number of bits per symbol, and the power allocated to each subchannel, the total bit rate of the system is the largest.
(4) Strong anti-fading ability by joint coding of subcarriers. OFDM technology itself has taken advantage of the frequency diversity of the channel. The performance of the system can be further improved by coding each channel.
(5) A fast algorithm based on discrete Fourier transform (DFT). OFDM adopts IFFT and FFT to achieve modulation and demodulation for the implementation of DSP which is easy to realize.

The basic idea of OFDM is to transform the high-speed serial data into multipath parallel data and modulate to each subcarrier for transmission. The parallel transmission technology greatly expands the pulse width of the symbol and improves the performance of anti-multipath fading. Orthogonal signals can be separated by using relevant techniques at the receiving end, which can reduce subcarriers interfering with the ISI. The signal bandwidth of each subcarrier is less than the associated bandwidth of the channel, so each subcarrier can be seen as flat decline, thus eliminating the inter-symbol interference. In the traditional method of frequency division multiplexing, each subcarrier frequency spectrum does not overlap each other and needs to use a lot of send filters and receive filters, thus greatly increasing the complexity and cost of the system. At the same time, in order to reduce the cross talk between subcarriers, the frequency spacing between the subcarriers must be maintained, which can reduce the frequency utilization of the system. The modern OFDM system adopts digital signal processing technology, and the generation and reception of each subcarrier are completed by digital signal processing algorithm, which greatly simplifies the structure of the system. In order to improve the spectrum efficiency at the same time, it is necessary to make each subcarrier frequency spectrum overlapping, but these spectrums in the whole symbol period satisfy orthogonality, ensuring that the receiver cannot recover the signal distortion.

The flow of OFDM generation and detection is shown in Fig. 8.15. The transmitter includes QAM mapping, transform from serial to parallel, IDFT, add CP, and transform from parallel to serial. The flow in receiver is just the opposite process to transmitter. At the transmitter, the information sequence is strung and converted to N parallel symbols and modulated separately on each branch. After the modulation, the parallel symbol is transformed by the fast Fourier transform (FFT) into a set of N different subcarriers and then combined with the protection interval. So, the OFDM signal is generated. The generated OFDM signal is amplified by the power amplifier, and then the signal works in the LED workspace through the dc offset. The signal is transmitted through the LED to the light intensity signal. At the receiving end, OFDM modulation signal is received by a photodiode (PD) which converts light intensity signal into a current signal. After the OFDM demodulation process, the original signal is restored. Cyclic prefixes are used to avoid the delay caused by multipath interference.

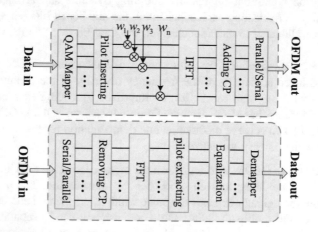

Fig. 8.15 Flowchart of generation and detection of OFDM signals

OFDM modulation has been widely used in visible light communication, including off-line systems and real-time systems. However, OFDM also has some disadvantages, including two main points: large PAPR; sensitivity to frequency.

We adopt OFDM modulation technology, combined with advanced pre-equilibrium and post-equilibrium algorithm, realizing the experiment of 875 Mbit/s visible bidirectional transmission system. The experimental block diagram is shown in Fig. 8.16.

In the experiment, OFDM signal generated by the arbitrary waveform generator (AWG) is, respectively, through a low-pass filter (LPF), the amplifier (EA), and offset the tree (bias tee) and then modulated to the different color chip RGB-LED. After free-space transfer, three wavelengths of light are separated by the filter and are received by the detector at the receiving end. Then, the equilibrium and demodulation algorithm are processed. The downlink uses red and green wavelengths, while the uplink adopts the blue wavelength. The experiment implemented the downlink 575 Mb/s, uplink 300 Mb/s full-duplex VLC transmission, and the transmission distance is 66 cm [15].

The frequency response of RGB-LED and the signal spectrum of red, green, and blue at three wavelengths are shown in Fig. 8.17. It can be easily found that the frequency of RGB-LED is in obvious decline at high frequency, and its 20 dB

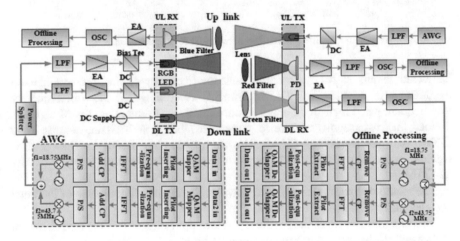

Fig. 8.16 The 875 Mbit/s two-way VLC transmission system based on OFDM modulation

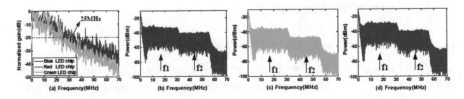

Fig. 8.17 Frequency response of RGB-LED and the signal spectrum at the red, green, and blue wavelengths

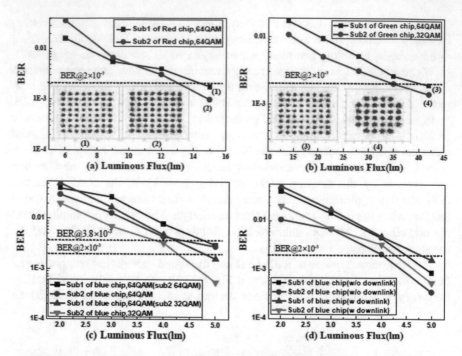

Fig. 8.18 Curve diagram of the relationship between BER and illuminance

bandwidth is only 25 MHz. To compensate the frequency decline, pre-equalization technology is adopted in the experiment. According to the difference of low-frequency and high-frequency response at the same time, the OFDM signal is divided into two segments. The low frequency has a high signal-to-noise ratio (SNR) and so adopts 64-QAM. However, the high frequency has a low signal-to-noise ratio (SNR) due to the high-frequency decline, thus using low-order 32-QAM.

The relationship between the three wavelengths and the illuminance of the three wavelengths are shown in Fig. 8.18.

This experiment proves the feasibility of the application of OFDM modulation in high-speed visible optical communication system.

8.1.3.2   ACO-OFDM Modulation

(1) The Principle of ACO-OFDM Modulation

The OFDM signal is a bipolar signal, but for the traditional intensity modulation–direct detection optical communication system, the bipolar signal cannot be used, because the light intensity cannot be negative. Therefore, for optical OFDM, a common solution is to add a DC bias on the OFDM signal in the electrical-to-optical signal conversion, which converts the bipolar signals into unipolar signals that can be transmitted through the optical channel.

But the OFDM with a DC bias increases the average optical power of a transmitted optical signal and reduces the modulation depth as well, leading to the low power efficiency of the system. We can now use a new way called ACO-OFDM to solve the problem of the dual polarity OFDM signal. The ACO-OFDM means artificially cutting off the negative part of the original OFDM signal and so only the positive part remains.

Suppose that the time domain sampling of the baseband OFDM symbol, with no circulation protection prefix, can be expressed as:

$$x(k) = \frac{1}{N} \sum_{m=0}^{N=1} X_m \exp\left(\frac{j2\pi km}{N}\right), \quad 0 \le k \le N - 1 \tag{8.5}$$

Here, $x(k)$ represents the time domain sampling value sequence, k represents the time sampling sequence index, N represents the subcarrier number, and X_m represents the frequency domain data symbols after mapping the modulation to the m subcarrier.

When X has the Hermitian symmetry structure, namely:

$$\begin{aligned} X &= (X_0, X_1, X_2, \ldots X_m, \ldots, X_{N-1}) \\ &= (X_0, X_1, X_2, \ldots X_{N/2-2}, X_{N/2-1}, X_{N/2-1}^*, \ldots, X_1^*) \end{aligned} \tag{8.6}$$

Here, X^* means the conjugate of X, and the OFDM signal after the IFFT is real. In this case, choose the odd subcarrier to modulate the data, and at the same time, set the even carrier modulation data to zero. Then, X can be expressed as:

$$X = (0, X_1, 0, \ldots X_{N/2-1}, 0, X_{N/2-1}^*, \ldots, 0, X_1^*) \tag{8.7}$$

When X satisfies the structure above, the OFDM signal after the IFFT is real, and then the negative part is set to zero prior to transmission. Next, the amplitude of the odd bits of transmitted information becomes half of the original after the demodulation, and the other information is not affected. Such a method that satisfies the real single polarity is called asymmetric chopping (ACO).

Next, we will introduce the principle of why the odd bit of the OFDM signal, after the asymmetric chopping transmission, is not affected by clipping noise. If only the odd bit of the carrier is not zero, its $m + 1$ subcarrier is expressed as:

$$x(m, k) = \frac{1}{N} X_m \exp\left(\frac{j2\pi km}{N}\right), \quad 0 \le k \le N - 1 \tag{8.8}$$

where $x(m, k)$ represents the $k + 1$ time domain sampling values of the $m + 1$ subcarrier. When m is an odd number, we can get that:

$$x(m, k + N/2) = \frac{1}{N} X_m \exp\left(\frac{j2\pi(k + N/2)m}{N}\right)$$

$$= \frac{1}{N} X_m \exp\left(\frac{j2\pi(k + N/2)m}{N}\right) \exp(j\pi m)$$

$$= -\frac{1}{N} X_m \exp\left(\frac{j2\pi km}{N}\right) \qquad (8.9)$$

$$= -x(m, k), \quad 0 \leq k \leq N - 1$$

According to the superposition principle, the sampling sequence of OFDM symbols is:

$$x(m, k + N/2) = -x(m, k), \quad 0 \leq k \leq N/2 - 1 \qquad (8.10)$$

To demodulate the data X_m, we can make an FFT transformation to the received $x(k)$:

$$X_m = \frac{1}{N} \sum_{k=0}^{N-1} x(k) \exp\left(\frac{-j2\pi km}{N}\right), \quad 0 \leq m \leq N - 1 \qquad (8.11)$$

We should handle $x(k) \leq 0$ and $x(k) > 0$ separately, as the following deformation shows:

$$X(m) = \frac{1}{N} \sum_{k=0}^{N-1} x(k) \exp\left(\frac{-j2\pi km}{N}\right)$$

$$= \frac{1}{N} \sum_{\substack{k=0 \\ x(k) \geq 0}}^{N/2-1} \left[x(k) \exp\left(\frac{-j2\pi km}{N}\right) + x(k + N/2) \exp\left(\frac{-j2\pi(k + N/2)m}{N}\right) \right]$$

$$+ \frac{1}{N} \sum_{\substack{k=0 \\ x(k) > 0}}^{N/2-1} \left[x(k) \exp\left(\frac{-j2\pi km}{N}\right) + x(k + N/2) \exp\left(\frac{-j2\pi(k + N/2)m}{N}\right) \right]$$

$$= \frac{1}{N} \sum_{\substack{k=0 \\ x(k) \geq 0}}^{N/2-1} \left[x(k) \exp\left(\frac{-j2\pi km}{N}\right) + x(k + N/2) \exp\left(\left(\frac{-j2\pi km}{N}\right) \exp(-j\pi m)\right) \right]$$

$$+ \frac{1}{N} \sum_{\substack{k=0 \\ x(k) > 0}}^{N/2-1} \left[x(k) \exp\left(\frac{-j2\pi km}{N}\right) + x(k + N/2) \exp\left(\frac{-j2\pi km}{N}\right) \exp(-j\pi m) \right]$$

$$0 \leq m \leq N - 1$$

$$(8.12)$$

Because only odd bits in the frequency domain data structures are nonzero, $e^{-jm} = (-1)^m = -1$, with Eq. (8.10), we can get:

$$X(m) = \frac{2}{N} \sum_{\substack{k=0 \\ x(k) \geq 0}}^{N/2-1} x(k) \exp\left(\frac{-j2\pi km}{N}\right)$$

$$+ \frac{2}{N} \sum_{\substack{k=0 \\ x(k) > 0}}^{N/2-1} x(k) \exp\left(\frac{-j2\pi km}{N}\right), \quad 0 \leq m \leq N-1 \tag{8.13}$$

If we have to do asymmetric clipping to the OFDM signal, as seen in Eq. (8.13), only the first one of the first plus symbol and the second one of the second plus symbol are not equal to zero. Thus, it can be expressed as

$$X_{m\_clip} = \sum_{k=0}^{N/2-1} x_{m\_clip} \exp\left(\frac{-j2\pi km}{N}\right)$$

$$= \frac{X_m}{2} \tag{8.14}$$

Among them, x_{m_clip} is the time domain sampling sequence after the asymmetric clipping before the receiver and X_{m_clip} is the demodulation data. From the equation, we can see the demodulation results: The useful information is on the odd bit of the carrier, and besides that its amplitude changes into half of the original, there was no other information effected by the asymmetric chopping.

(2) Theory of the ACO-OFDM Modulation System

The principles of ACO-OFDM modulation are similar to that of OFDM. As shown in Fig. 8.19, it is the principle diagram of the ACO-OFDM modulation system.

The principles of the ACO-OFDM modulation system are as follows:

(1) Make QAM constellation mapping for the input binary sequence. If using 16-QAM (four-order quadrature amplitude modulation), the constellation corresponding to the transmitted information is $\{\pm1, \pm j\}$, $\{\pm3, \pm3j\}$.
(2) Add the conjugated complex data to the complex signal mapped by constellations to constitute Hermitian symmetry, and then deliver it to the IFFT module after serial-to-parallel conversion.
(3) Perform ACO to the bipolar real sampling values after IFFT, and make it into a single polarity positive signal by restricting the amplitude at the zero-value points, that is:

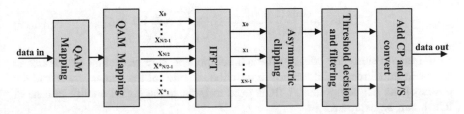

Fig. 8.19 Principle diagram of the ACO-OFDM modulation system

$$x' = \begin{cases} x, & \text{if} \quad 0 \leq x' \geq A' \\ 0, & \text{if} \quad x' > 0 \end{cases} \tag{8.15}$$

And for the positive real number signal x', set the threshold value A and the judgment criteria is as follows:

$$y' = \begin{cases} x', & \text{if} \quad 0 \leq x' \geq A' \\ 0, & \text{if} \quad x' > 0 \end{cases} \tag{8.16}$$

Next, by performing frequency domain filtering after the threshold decision, we can then get the baseband OFDM signal. The frequency domain filtering can alleviate the out-of-band radiation energy, inhibiting the out-of-band noise.

(4) Add CP to the baseband OFDM positive real signal after the asymmetric chopping and filtering, and finally, the output signal is obtained after the serial-to-parallel conversion.

(3) The ACO-OFDM Simulation Performance

According to the analysis of the ACO-OFDM principles above, we can further simulate the performance of the ACO-OFDM, to illustrate that the ACO-OFDM modulation method has a good resistance to ICI.

After the ACO-OFDM signal passes through the channel, the time domain signal received at the receiver can be expressed as:

$$y(n) = x(n)e^{\frac{j2\pi n \varepsilon}{N}} + w(n) \tag{8.17}$$

where ε means the normalized frequency offset, $\varepsilon = \Delta f \cdot NT_S$, Δf is the Doppler frequency shift, T_S is the subcarrier symbol period, and $w(n)$ is the white Gaussian noise channel additive. After FFT at the receiver, the frequency domain signal is:

$$Y(k) = \sum_{l=0}^{N-1} y(n)e^{-\frac{j2\pi kn}{N}} = X(k)S(0) + \sum_{\substack{l=0 \\ l \neq k}}^{N-1} X(l)S(l-k) + W(k) \tag{8.18}$$

where $W(k)$ is the FFT transformation of $w(n)$ and $\sum_{\substack{l=0 \\ l \neq k}}^{N-1} X(l)S(l-k)$ is the impact of the K subcarrier brought about by ICI.

This is so because:

$$S(l) = \frac{\sin[\pi(l+\varepsilon)]}{N\sin[\pi(l+\varepsilon)/N]} \exp\left[j\pi\left(1 - \frac{1}{N}\right)(l+\varepsilon)\right] \tag{8.19}$$

In order to study the effect of the ICI component, here we define the parameter carrier-to-interference (CIR) ratio, to describe the ratio of the signal power and the power of the ICI component. CIR is used to indicate the signal quality in case of ICI. It can be expressed as:

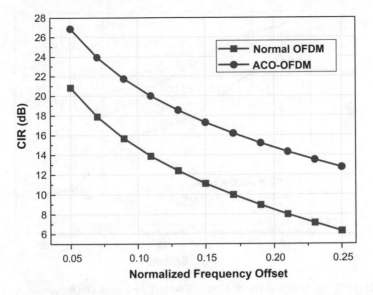

Fig. 8.20 CIR versus the normalized frequency offset curve for ACO-OFDM and normal OFDM

$$\text{CIR} = \frac{E\left[|C(k)|^2\right]}{E\left[|\text{ICI}(k)|^2\right]} = \frac{E\left[|X(k)S(0)|^2\right]}{E\left[|X(l)|^2\right] \cdot \sum_{\substack{l=0 \\ l \neq k}}^{N-1} |S(l-k)|^2} = \frac{|S(0)|^2}{\sum_{l=1}^{N-1} |S(l)|^2} \quad (8.20)$$

Based on the above formula, we made simulations on ACO-OFDM and normal OFDM. The CIR versus the normalized frequency offset ε curve is shown in Fig. 8.20.

As seen in Fig. 8.20, because in the ACO-OFDM modulation the subcarriers are vacated to 0, it can then reduce the ICI. Compared with normal OFDM, under the same frequency offset condition, the CIR of ACO-OFDM has been enhanced by 6 dB.

Next, we combine ACO-OFDM with the visible light communication system and carry on the performance simulation. We use $H(nw_0) = (e^{-w_0/w_b})^n$ as the VLC channel frequency response, where w_0 is the minimum frequency of OFDM subcarrier and w_b is the matching coefficient. We set $w_b = 2\pi \times 21 \times e^6$ rad/s in the simulation to simulate the actual visible light communication channel. Then, the pre-equalized ACO-OFDM signal is transmitted through the fitting channel and demodulated in the receiver. The BER versus SNR and normalized spectrum curves are shown in Figs. 8.21 and 8.22, respectively.

From Fig. 8.22, we can see that compared with the normal OFDM modulation, when under the same BER, the SNR, which ACO-OFDM requires, decreases by 3 dB. Meanwhile, under the same influence of the frequency offset, the BER performance of ACO-OFDM improves by 5 dB.

The above simulation results further demonstrate the advantages of ACO-OFDM in resistance to ICI and the system frequency offset.

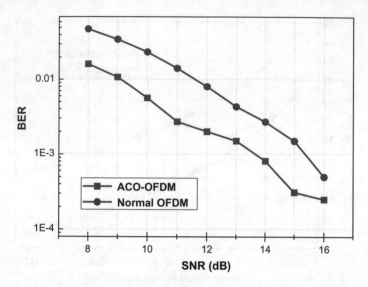

Fig. 8.21 BER versus SNR curve of ACO-OFDM and the normal OFDM system

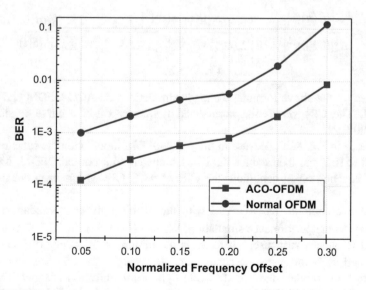

Fig. 8.22 BER versus the normalized frequency curve of ACO-OFDM and the normal OFDM system

(4) **The Advantages and Disadvantages of ACO-OFDM Modulation**

Besides the advantages of OFDM, ACO-OFDM also has other unique advantages as follows:

(1) The transceiver structure is simple. Due to the Hermitian symmetry structure, OFDM signals are real rather than complex. In addition, they no longer need to be divided into a real part and an imaginary part for the orthogonal modulation–demodulation process, which reduces the system cost.
(2) The efficiency of optical power is high. Because it directly sets the negative polarity part of the signal to zero, it is not necessary to add a DC bias, which can improve the power efficiency. In addition, when applied to optical wireless communication, it is more conducive to eye safety.
(3) The modulation depth is high. Because ACO-OFDM modulation does not need a DC bias, we can directly offset the OFDM signal at zero, which can improve the modulation depth of the system and increase the modulation efficiency.
(4) The influence caused by nonlinearity is small. The main reason for optical fiber nonlinearity is the overlarge light power. Because AC technology can reduce the required optical power, the nonlinear effect is relatively reduced, which no longer makes dispersion the main reason that restricts the system's performance.

Any performance improvements have their costs, and the main drawbacks of ACO-OFDM system are as follows:

(1) The spectrum efficiency is relatively lower. Because the system adopts a Hermitian symmetrical structure and abandons the modulation data of the even carriers, only the data of N/4 points is independent of each other in N points IFFT.
(2) A new timing algorithm is needed. Because it abandons the negative polarity part of the signal, the performance of original classical timing methods, such as Schmidl timing and Park timing, seriously decreases. Thus, it needs a new timing algorithm.

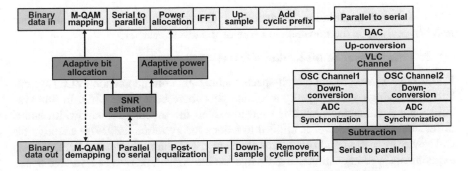

Fig. 8.23 Block diagram of implementing bit and power allocation with OFDM

8.1.4   Bit-Loading OFDM

(1) The Principle of bit-loading OFDM Modulation

The diagram of implementing bit and power allocation with OFDM is shown in Fig. 8.23. In the implementation, the OFDM transmitter consists of M-ary QAM (M-QAM, 1 for BPSK, 2 for QPSK, 3 for 8-QAM, 4 for 16-QAM, etc.) modulation based on adaptive bit allocation, serial-to-parallel conversion, power allocation based on adaptive power allocation, inverse fast Fourier transform (IFFT), up-sample, adding cyclic prefix (CP), parallel-to-serial conversion, digital-to-analog converter (DAC), and up-conversion. DAC is used to transform the time domain digital data to time domain analog data. Up-conversion is performed to up-convert the signal to the transmission frequency. The OFDM signal is loaded to an AWG as the input of the VLC system and goes through the VLC channel. At the receiver, the final differential output signals are captured by channel 1 and channel 2 of OSC. An off-line digital signal processing program by MATLAB is used to demodulate the OFDM signal. First, differential signal channel 1 subtracts channel 2 after synchronizations. Then, the resulted signal is demodulated by the following steps, including removing CP, serial-to-parallel conversion, down-sample, fast Fourier transform (FFT), post-equalization, and M-QAM decoding. The post-equalization method is zero-forcing equalization by a training sequence. The M-QAM decoding needs the bit allocation information of every subcarrier. At last, we calculate the BER by comparing the original binary data and recovered binary data. Other detailed parameters of the generated OFDM signals include: subcarrier number = 256, up-sampling factor = 3. The data rate includes the 3.03% CP, 2% training sequence, and 7% forward error correction (FEC) overhead.

Before applying bit and power allocation, the signal-to-noise ratio (SNR) of the VLC channel is estimated through error vector magnitude (EVM) method using BPSK-OFDM [7]. The total data rate can be calculated as

$$R = \frac{B\left(\sum_{k=1}^{N} \log_2 M_k\right)}{N} \tag{8.21}$$

where B is the modulation bandwidth of the system, N is the total subcarrier number, and M_k is the constellation size of the kth subcarrier.

(2) The experiment of bit-loading OFDM

In this section, we present a high-speed visible light communication (VLC) system based on a single commercially available phosphorescent white LED. In this system, a pre-equalization circuit is used to extend the modulation bandwidth and a differential output receiver is utilized to reduce the system noise. With adaptive bit and power allocation and orthogonal frequency division multiplexing (OFDM), we experimentally demonstrated a 2.0-Gb/s visible light link over 1.5 m free-space transmission, and the BER is under pre-forward error correction (pre-FEC) limit of

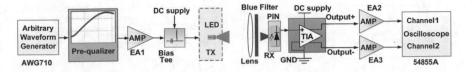

Fig. 8.24 Experimental setup of VLC system

3.8×10^{-3}. To the best of our knowledge, this is the highest white light VLC data rate using a single phosphorescent white LED.

The experimental setup of VLC data transmission system using the pre-equalizer is shown in Fig. 8.24. In this scheme, the original driving signal is from AWG (Tektronix AWG710) and pre-equalized by the amplitude equalizer. After amplified by EA1 (Minicircuits, gain: typical 25 dB, minimum output power at 1 dB compression: 22 dBm, and −3 dB bandwidth: 500 MHz), the amplified signal is combined with direct current (DC) using bias tee and applied to a single commercially available phosphorescent white LED (OSRAM, LCWCRDP.EC) acting as the optical transmitter. In order to focus a high proportion of light on the PIN receiver, a lens (diameter: 55 mm, focus length: 18 mm) is placed before the receiver. The blue filter is placed in front of the PIN photodiode to filter out the slow-responding phosphor component. The blue filter has a very high transmittance average 97.5% from 430 to 485 nm in the blue signal range and a very wide stop-band from 500 to 1050 nm [8]. A low-cost commercial PIN photodiode (Hamamatsu S10784, effective photosensitive area: 7 mm^2, and 0.45 A/W sensitivity with −3 dB bandwidth of 300 MHz at 660 nm) is used to convert optical signal to electrical signal. Then, the electrical signal is amplified by a differential output trans-impedance amplifier (TIA) designed with MAX3665 (gain of 8 kΩ and −3 dB bandwidth of 470 MHz). The differential outputs of TIA are, respectively, amplified by EA2 and EA3 (Minicircuits, gain: typical 25 dB, minimum output power at 1 dB compression: 22 dBm, and −3 dB bandwidth: 500 MHz). The output signals of EA2 and EA3 are recorded simultaneously by channel 1 and channel 2 of a real-time digital oscilloscope (OSC, Agilent 54855A).

(3) The results and discussions

To maximize the transmission data rate of the VLC system, we have designed several different kinds with different parameters, including single and cascaded pre-equalizers. We find the most suitable pre-equalizer using for the bit and power allocation OFDM. The parameters of the best pre-equalizer used in this paper are $R1 = 499\ \Omega$, $R2 = R3 = 49.9\ \Omega$, $R4 = 5\ \Omega$, $C1 = C2 = 8.5$ pF, $L1 = L2 = 22$ nH. The forward transmission gains of the differential outputs are measured by the vector network analyzer (VNA, Agilent, N5230C) operating from 10 MHz to 40 GHz, and the output power of VNA is fixed at −25 dBm, relatively small to avoid saturation distortion. The distance between the transmitter LED and receiver PIN is 150 cm. In Fig. 8.25, using the blue filter, we show the measured forward transmission gains of the differential outputs with pre-equalizer and without

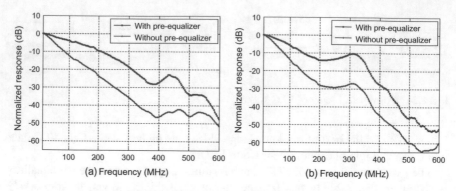

Fig. 8.25 Measured forward transmission gains **a** channel 1 and **b** channel 2

pre-equalizer in VLC system. Using blue filter, the −3 dB bandwidths of channel 1 and channel 2 are both 28 MHz (from 10 to 38 MHz), respectively, and improved to 66 MHz (from 10 to 76 MHz) and 55 MHz (from 10 to 55 MHz) using the pre-equalizer. With lower frequencies attenuated by using the passive equalizer, the transmit power with equalizer is larger than that without equalizer to achieve the best working condition [9].

In the data transmission experiments, the distance is 150 cm between the transmitter LED and receiver PIN. The sampling rates of the AWG and OSC are fixed at 1.8G Sample/s and 2G Sample/s, respectively. The modulation bandwidth is 600 MHz from DC to 600 MHz. SNR estimation is performed using

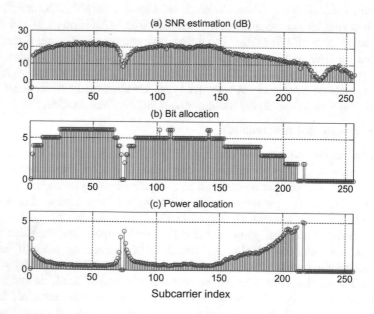

Fig. 8.26 a SNR estimation, **b** bit, and **c** power allocation scheme of 2.28 Gb/s at 150 cm

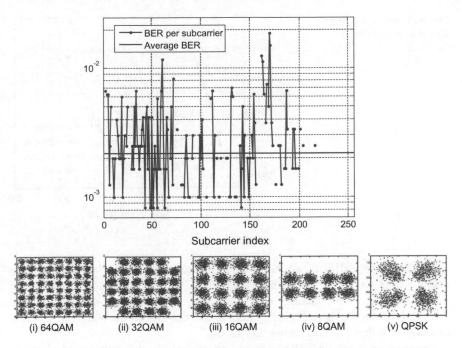

Fig. 8.27 Measured BER results on individual subcarriers and the constellations of 2.28 Gb/s at 150 cm

BPSK-OFDM, and all the individual subcarrier has equal power. After exhaustive experiments by varying the signal driving peak-to-peak voltage (V_{pp}) of AWG and bias current (I_{bias}) of the phosphorescent white LED, we obtain the optimal biasing point to be $V_{pp} = 1.2$ V and $I_{bias} = 426$ mA. The illuminance is about 651 lx measured before the lens at the distance of 150 cm using the light meter (Fluke, 941, Light Illuminance Meter Tester). The bit and power allocation scheme utilized in this paper are based on the work of [10].

In Fig. 8.26a, we show the estimated SNR of VLC channel versus subcarrier index. Based on the estimated SNR, the adaptive bit allocation algorithm can adaptively allocate the modulation order (bits/symbol) to different subcarriers, that is, modulation order higher with the better SNR, so the bandwidth can be fully employed and the total data rate will be improved. Fixing the transmission data rate at 2.28 Gb/s (average bit per subcarrier: 3.8 bits/symbol), the bit and power allocation of different subcarriers are presented in Fig. 8.26b, c, respectively. In Fig. 8.26b, 46 subcarriers are omitted because of the lower SNR and the frequency of the last subcarrier is 520.8 MHz. The assigned maximum modulation order is 6 bits/symbol (64-QAM) and the minimal is 2 bits/symbol (QPSK). The power allocation is illustrated in Fig. 8.26c, and the modulation orders have been considered when implementing the power loading algorithm. In Fig. 8.27, we present the BER result as a function of the subcarrier index. It can be found that some subcarriers have higher BER since the channel has changed slightly at different

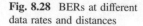

Fig. 8.28 BERs at different data rates and distances

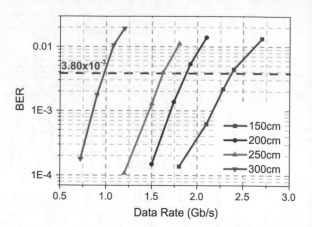

time. The average BER of all subcarriers (excluding the subcarriers without allocating bits) is 2.17×10^{-3} and under the pre-FEC limit of 3.8×10^{-3}. Considering the 3.0% CP, 2% training sequence, and 7% forward error correction (FEC) overhead, the data rate is 2.0 Gb/s at 150 cm The constellations of different modulation orders are also shown in Fig. 8.27.

In Fig. 8.28, we show BER results as a function of the data rate with the distance from 150 to 300 cm. The illumination levels range (approximately) from 651 lx (150 cm) to 134 lx (300 cm). The BER results increase with higher data rate and longer transmission distance. When the transmission distance becomes longer, the received optical power at receiver will decrease and the SNR of the VLC system deteriorates, so the data rate should be reduced to meet the requirement of the pre-FEC limit. The total data rates are, respectively, 2.28, 1.74, 1.50, and 0.90-Gb/s over the distance of 150, 200, 250, and 300 cm, and considering the overhead, the data rates will be 2.00, 1.53, 1.32, and 0.79-Gb/s. To date, it is the highest data rate ever achieved by employing a single commercially available phosphor-based white LED in VLC systems reported.

In this section, we demonstrate a very high-speed VLC system based on a single commercially available phosphorescent white LED. In this VLC system, a pre-equalization circuit is used to extend the modulation bandwidth and a low-cost differential output PIN receiver is utilized to reduce the system noise and increase data rate. Combining OFDM and adaptive bit and power allocation algorithm, we successfully realize a 2.0-Gb/s VLC link over 1.5 m free-space transmission and 0.79-Gb/s data rate over 3.0 m, with the BER results under pre-FEC limit of 3.8×10^{-3}. This demonstration is conducted at a practical distance. To the best of our knowledge, this is the fastest white light VLC data rate transmission using a single phosphorescent white LED ever achieved.

8.2 Multi-user Access and Bidirectional VLC System

So far, VLC investigations have been mostly focused on point-to-point, unidirectional demonstration based on various techniques such as orthogonal frequency division multiplexing (OFDM), multiple-input multiple-output (MIMO), and post-equalization. Bidirectional VLC transmission is still a challenge due to the lack of good resolution of the uplink of an indoor VLC system. A retro-reflecting transceiver has been proposed; however, it cannot achieve a high modulation rate because the reflection is realized in a pure physical way. Employing LED VLC for bidirectional transmission via time division duplex has been demonstrated; however, it needs precise synchronization and the data rate of uplink and downlink is only 2.5 Mb/s.

In order to achieve the high-speed access services of indoor VLC system, we need to solve the problems of LED-based multi-user access and bidirectional communication. In this section, we introduce the MISO multi-user access techniques and experiments of bidirectional VLC system

8.2.1 The Multiple-Input and Single-Output System

The principle diagram of a subcarrier multiplexed $N \times 1$ MISO-OFDM VLC system is shown in Fig. 8.29. The input binary sequences are first modulated in

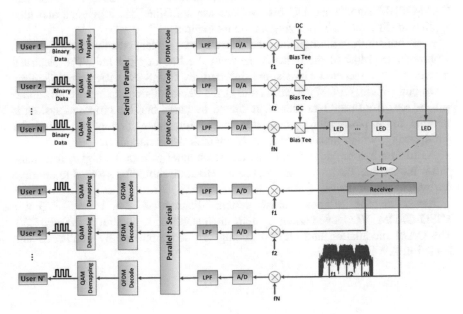

Fig. 8.29 Diagram of the subcarrier multiplexed $N \times 1$ OFDM-MISO VLC system

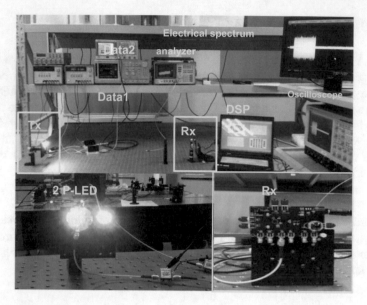

Fig. 8.30 MISO experimental setup

QPSK or QAM format and then passed to the OFDM encoder. Then, the QAM-OFDM signals are up-converted to different subcarriers and added up. Subsequently, the multiplexed QAM-OFDM signals coming from the AWG are filtered by a low-pass filter (LPF) and amplified by an EA. The electrical QAM-OFDM signals and DC bias voltage are combined via a bias tee and then applied to different LEDs serving as the transmitter.

After transmitting through the indoor free space and lens, the light will focus on a photodiode. Most of the lights come from the line of sight link. The electrical spectra of the received signals are inserted in Fig. 8.29. After amplification, down-conversion, OFDM decoding, and QAM demodulation, the recovered signals will be sent from user 1 to user N. It should be noted that the proposed system is suitable for a positioning system due to the implementation of the subcarrier multiplexing, which makes it easy to distinguish the transmitters.

As for the indoor white light channel, the channel gain of the higher frequency part is higher than the lower one. So, pre-equalization should be adopted to improve the system performance.

Based on the aforementioned system diagram, a 2×1 and a 3×1 MISO-OFDM VLC experiment system are established, as shown in Fig. 8.30. The QAM modulation and OFDM coding are accomplished by MATLAB and loaded in AWG.

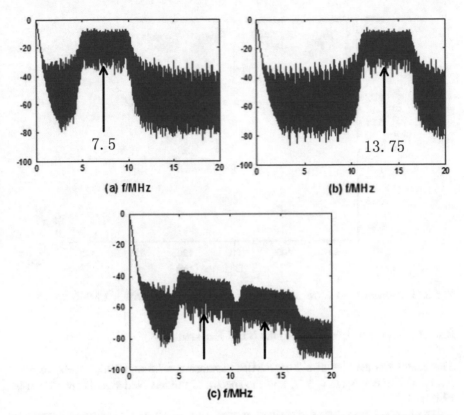

Fig. 8.31 Electrical spectra of the received signals **a** sub1 (7.5 MHz), **b** sub2 (13.75 MHz), **c** sub1, and sub2

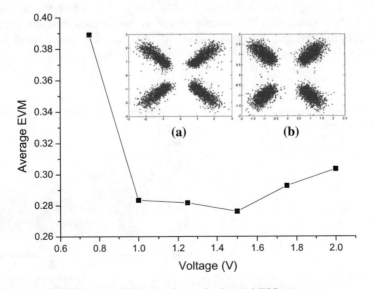

Fig. 8.32 Average EVM versus the input voltage of a 2 × 1 MISO system

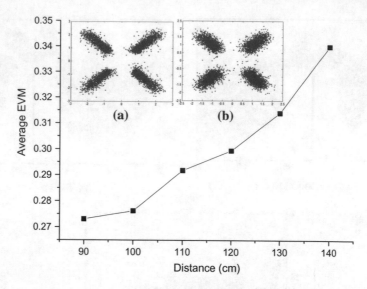

Fig. 8.33 Average 2 × 1 EVM versus the transmission distance of a 2 × 1 MISO system

8.2.1.1   The 2 × 1 MISO-OFDM VLC Experiment

The center frequencies of a 2 × 1 MISO system are 7.5 and 13.75 MHz, respectively. The bias voltage is 3 V, and the spectra of the received signals are shown in Fig. 8.31.

The transmission distance is fixed at 1 m, and the EVM versus input voltage is measured and shown in Fig. 8.32. The performance is not good when the voltage is too large or small. In addition, the optimal input voltage is from 1 to 1.5 V. The constellation of sub1 (7.5 MHz) and sub2 (13.75 MHz) is inserted as shown in Fig. 8.32a, b, respectively. The average EVM versus the transmission distance is measured in the case that the input voltage is 1.5 V, and the results are depicted in Fig. 8.33.

Fig. 8.34 Electrical spectra
of received signals

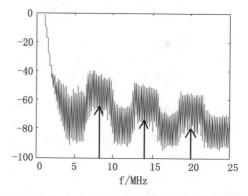

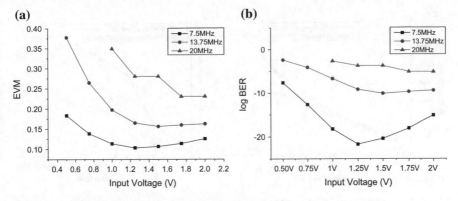

Fig. 8.35 a EVM of a 3 × 1 MISO system. **b** The BER of a 3 × 1 MISO system

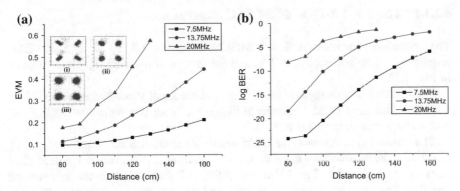

Fig. 8.36 a EVM distance of a 3 × 1 MISO system. **b** The BER versus the transmission distance of a 3 × 1 MISO system

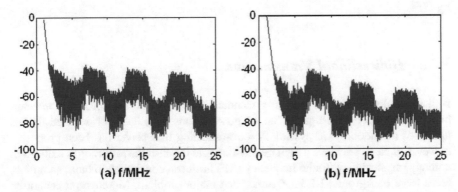

Fig. 8.37 Measured electrical spectra of received signals (**a**) with and (**b**) without pre-equalizations

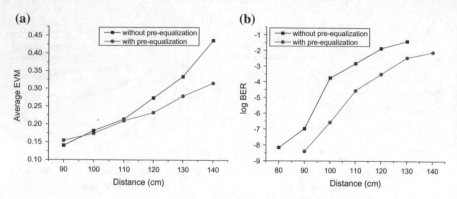

Fig. 8.38 a EVM and **b** the BER versus the transmission distance with and without pre-equalizations

8.2.1.2   The 3 × 1 MISO-OFDM VLC Experiment

The center frequencies of a 3 × 1 MISO system are 7.5, 13.75, and 20 MHz, respectively. The bias voltage is 3 V, and the spectra of received signals are shown in Fig. 8.34.

The transmission distance is set at 1 m, and the input voltages vary from 0.5 to 2 V. The measured EVM is shown in Fig. 8.35. From this figure, we realize the best voltage area is from 1.3 to 1.75 V.

The transmission distances of EVM versus BER are measured and depicted in Fig. 8.36. The bias voltage is set at 1.5 V. The inset (i), (ii), (iii) is the constellations of sub1 to sub3. The EVM and BER will degrade with the increase of distance. The electrical spectra both with and without pre-equalizations are measured and depicted in Fig. 8.37.

By adopting pre-equalizations, the transmission distance can be enhanced, as shown in Fig. 8.38. At a distance of 1.4 m, the EVM performance can be enhanced by 30%.

8.2.2   Bidirectional Transmission

Bidirectional transmission remains a technical challenge. So far, several approaches have already been investigated, but no general consensus has been reached for the uplink of an indoor VLC system. A retro-reflecting transceiver has been proposed; however, it cannot achieve a high modulation rate because the reflection is realized in a purely physical way. Radio frequency (RF) has been considered to build an uplink, but a large background RF interference is a major problem. Moreover, it cannot be used in some RF prohibited areas. In this section, we will introduce two bidirectional schemes, which are the time division duplex and the frequency division duplex.

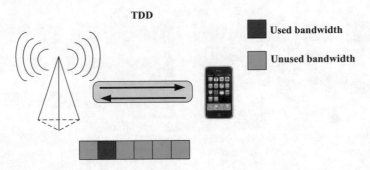

Fig. 8.39 Principle of TDD

8.2.2.1   Time Division Duplexing

In the time division duplexing (TDD)-based bidirectional transmission, the uplink and downlink are isolated by different time slots. In addition, they can work in the same frequency. The principle of TDD is shown in Fig. 8.39.

Liu et al. [11] have successfully realized bidirectional through TDD. In their demonstration, they adopt 5 × 8 LED arrays for the downlink and a single LED chip for the uplink. The synchronous signals are square signals with a duty ratio of 50%. OOK is employed in both the uplink and the downlink, and the data rate is 2.5 Mb/s.

8.2.2.2   Frequency Division Duplexing

In the frequency division duplexing (FDD)-based bidirectional transmission, the uplink and downlink work are in the same time slot but at a different frequency. The principle is shown in Fig. 8.40.

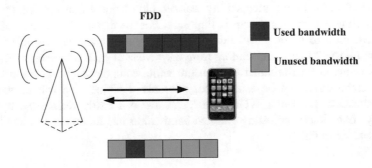

Fig. 8.40 Principle of FDD

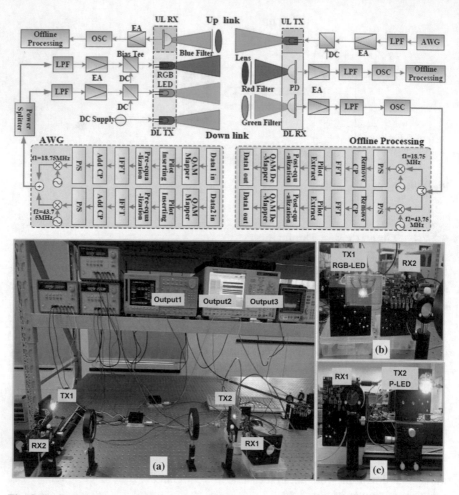

Fig. 8.41 Experiment principle and setup of FDD

Wang et al. [12] have successfully realized bidirectional transmission by FDD. The experimental setup is shown in Fig. 8.41. For the downlink, the red and green colors are used to carry useful information for communication, and in addition, there are two SCM channels starting from 6.25 MHz at each wavelength. The blue color is reserved for illumination to ensure white color lighting. For the uplink, a similar architecture can be established, but only the blue chip can be used for communication. By using WDM and SCM, we can easily realize the dynamic capacity allocation by adjusting different bandwidths and modulation orders for the uplink and downlink.

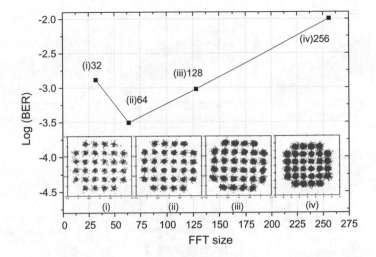

Fig. 8.42 Effect of FFT

Fig. 8.43 Measured BER versus input power of P-LED

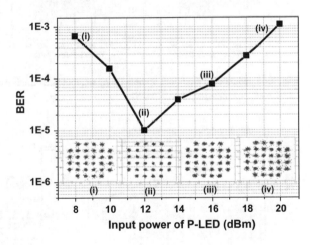

Regarding the QAM-OFDM modulation technique, the pre- and post-equalizations are employed in this system to maximize the transmission capacity. Prior to the demonstration, the effect of the FFT size has been measured. The FFT sizes of 32, 64, 128, and 256 are measured, and the results are illustrated in Fig. 8.42. From this figure, we discover that 64 has the best performance.

The nonlinearity effect introduced by the LED chips will also influence the system. The input power of a P-LED varies from 8 to 20 dBm with a 2-dB step. The results of this demonstration are presented in Fig. 8.43. As we can see, the optimal input power is 12 dBm. A lower input power will reduce the signal-to-noise ratio (SNR) and cause low modulation depth, while a higher one will cause nonlinearity and clipping.

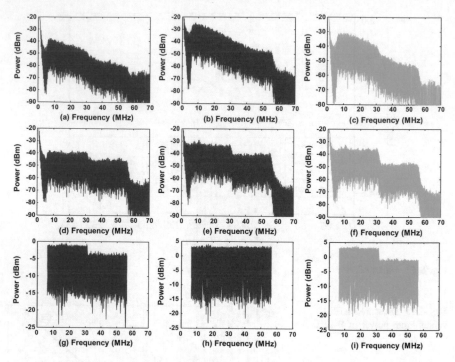

Fig. 8.44 Measured electrical spectra (**a–c**) without pre-equalization, (**d–f**) with pre-equalization, (**g–i**) with pre- and post-equalizations

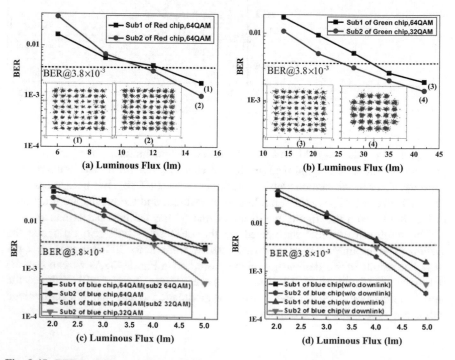

Fig. 8.45 BER performance of each color

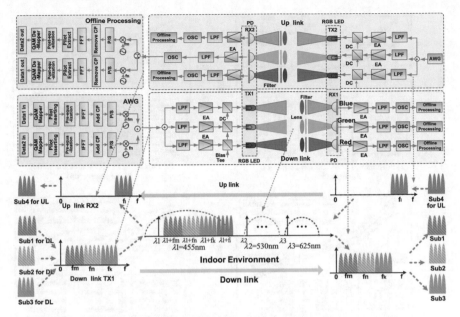

Fig. 8.46 Experimental setup of the WDM-SCM bidirectional transmission system

The multiplexed QAM-OFDM signals are generated by AWG720 and then carried on a LED. After free-space transmission, the optical signals are detected by the PIN and then captured by real-time OSC. The data rates of the red, green, and blue LED chips are 300, 275, and 300 Mbit/s, respectively.

The frequency characteristics of the electro-optical-electro channel are measured, and the bandwidths around the 20 dB point are all about 25 MHz. According to the channel knowledge, pre-equalization has been designed and applied. The electrical spectra of the received signals, with (w) and without (w/o) equalizations at each wavelength, are measured and depicted in Fig. 8.44. The BER performance of each color LED is measured and shown in Fig. 8.45a–c. The cross talk caused by the bidirectional transmission is also analyzed. As the uplink and downlink behave similarly, we choose the uplink for discussion. Figure 8.45d shows the BER performance of the uplink, both with and without the downlink. It can be easily seen that there is almost no degradation, which shows that the cross talk is quite low.

Compared to the previous scheme, this scheme, which is based on WDM and SCM, can provide a more flexible bandwidth allocation to obtain a higher transmission data rate. In this demonstration, each of the downlink wavelengths is divided into three subchannels, while the uplink has one subchannel. The bandwidths of all of the subchannels are 25 MHz, and the modulation orders are assigned according to the channel characteristics, which range from 16-QAM to 64-QAM. The experimental setup is shown in Fig. 8.46, and the results are depicted in Fig. 8.47. The data rates of the downlink and the uplink are 1.15 Gb/s and 300 Mb/s, respectively.

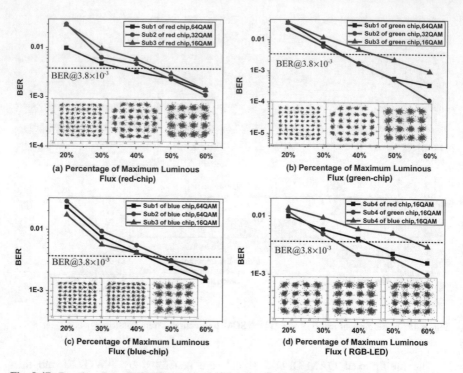

Fig. 8.47 Results of the WDM-SCM-based bidirectional transmission

8.3 VLC Multidimensional Multiplexing

In recent years, utilizing white light-emitting diodes (LEDs) for simultaneous illumination and visible light communication (VLC) has attracted more and more interest. The LED-based VLC system provides cost-effective, license-free, electromagnetic interference-free and secure communication link. However, there are still some bottleneck constraints to impede the development of VLC technology, one of which is the limited modulation bandwidth of LED. To break through the modulation bandwidth bottleneck, researchers have proposed a variety of technologies, such as high-order modulation formats, blue light filtering, pre-equalization, post-equalization, to improve the VLC system transmission rate.

In addition, it is an effective way to overcome the modulation bandwidth limitations by achieving multichannel parallel transmission, thus improving the capacity of VLC system [33]. In VLC system, various multiplexing methods can be adopted including wavelength division multiplexing (WDM), subcarrier

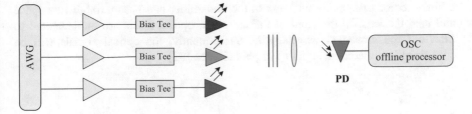

Fig. 8.48 Principle diagram of WDM

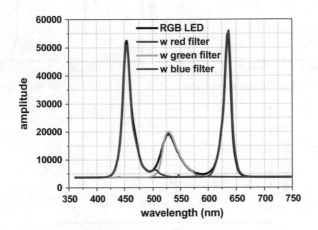

Fig. 8.49 Optical spectra of RGB-LED

multiplexing (SCM), and polarization division multiplexing (PDM).WDM refers to the signal modulated onto the visible light carrier of different wavelengths, then coupling into white light to transmit in free space. SCM refers to the signal modulated onto the LED subcarrier of different frequencies. PDM refers to two channels of information to be transmitted on the same carrier frequency by using waves of two orthogonal polarization states. In this section, we apply various multiplexing techniques to achieve high-speed VLC system.

8.3.1 Wavelength Division Multiplexing (WDM)

Wavelength division multiplexing technology utilizes multiple different wavelengths for data transmission. In a single RGB-LED, the red, green, and blue colors can each carry different information, and the overall data rate can be tripled. The schematic diagram of WDM is depicted in Fig. 8.48. Three different data streams are modulated on three different colors, and the lights emitted from the three different color LED chips will mix at the transmitter and pass through free space and then the lens. In the front of the receivers, optical filters are implemented to filter out the unwanted wavelengths. Subsequently, the optical signals will be conversed into electrical signals via photodiodes [32].

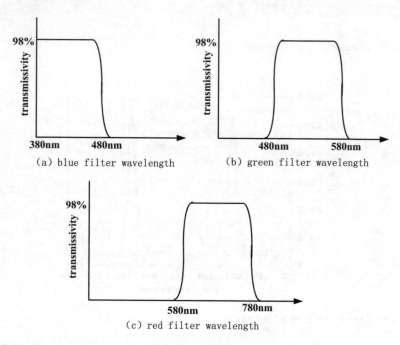

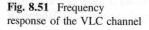

Fig. 8.50 Transmission curve of the R/G/B optical filters

Fig. 8.51 Frequency response of the VLC channel

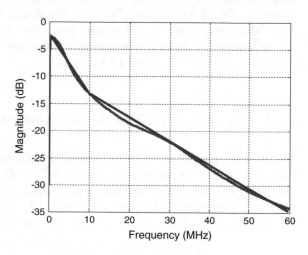

The optical spectra of RGB-LED before and after the optical filters are illustrated in Fig. 8.49. It can be easily seen that the three different colors can be separated via different optical filters.

The transmission curve of each optical filter is depicted in Fig. 8.50a–c.

Like the WDM technology in optical fiber communications, wavelength multiplexing and de-multiplexing devices (in the VLC system, these devices are optical filters) are essential components in the VLC system. The performance of the optical filters will affect the whole system. The main requirements for optical filters are a high transmittance for the designated wavelength, high absorptivity for the unwanted wavelength, and a relatively flat response of the transmission curve.

Cossu et al. [13] employed an RGB-LED (Cree PLCC6 Multichip LED), which emitted a luminous flux of 6 lm. In addition, the center wavelengths of red, green, and blue are 620, 520, and 470 nm, respectively. The aggregate transmission data rate can reach up to 3.4 Gb/s by employing WDM and OFDM, with the assistance of a bit and power loading algorithm.

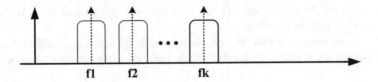

Fig. 8.52 Principle of SCM

Fig. 8.53 Sketch of SCM

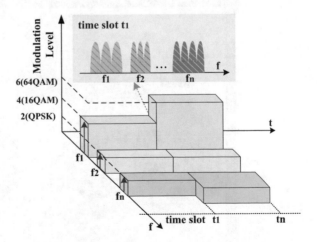

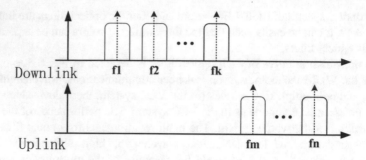

Fig. 8.54 Sketch of SCM in bidirectional transmission

8.3.2   *Subcarrier Multiplexing (SCM)*

Subcarrier multiplexing (SCM) can transmit different signals at one time. This flexible frequency allocation method makes it a promising candidate in bidirectional transmission and multiple access [34]. The frequency response of VLC is depicted in Fig. 8.51, which is not flat.

OFDM can overcome this uneven characteristic by adopting multiple subcarriers. However, its PAPR will increase with the increasing number of subcarriers.

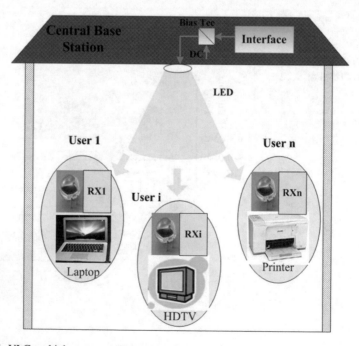

Fig. 8.55   VLC multiple access technique

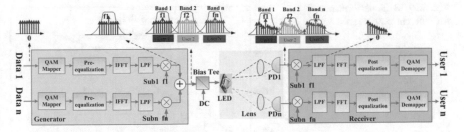

Fig. 8.56 Experimental principle of VLC multiple access

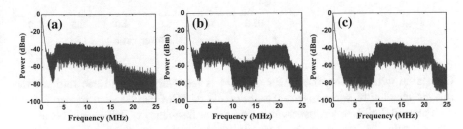

Fig. 8.57 Frequency spectra of different combinations

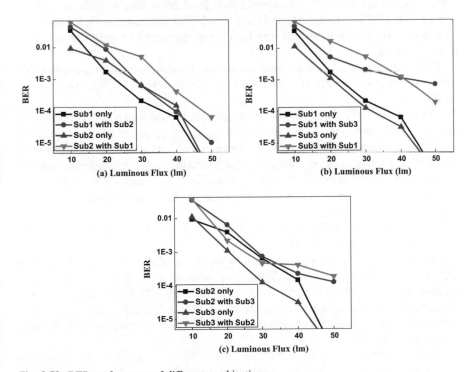

Fig. 8.58 BER performance of different combinations

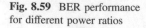

Fig. 8.59 BER performance
for different power ratios

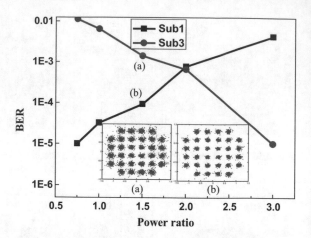

Compared with OFDM, the PAPR of SCM is relatively low and can mitigate the uneven channel to some extent. The principle of SCM can be seen in Fig. 8.52.

The modulation order, bandwidth, and center frequency of different subchannels can be adjusted according to different user's demands, which can be seen in Fig. 8.53. The available bandwidth can be divided into N subchannels, whose center frequencies are $f1, f2, \ldots fn$, respectively. In addition, their modulation formats can be OOK, QPSK, 16-QAM, 32-QAM, or a higher order modulation format. As this system is asynchronous, the bandwidth and modulation format can be adaptively adjusted.

SCM can be applied in both multiple access and bidirectional transmission due to its advantages. The sketch of its bidirectional transmission application is shown in Fig. 8.54.

The frequency allocation for downlink and uplink is decided by the demand. The sketch is illustrated in Fig. 8.54. Here, f_1, f_2, f_k, f_m and f_n are the center frequencies of different subcarriers. Different signals can be carried on these subcarriers. The modulation format, center frequencies, and bandwidth of each subchannel can be allocated according to the aforementioned principle. For example, the modulation format of downlink can be QAM-OFDM, while the data rate of uplink is relatively low, and it can be OOK.

The principle of multiple access is illustrated in Fig. 8.55, and its experiment principle is shown in Fig. 8.56.

Multiple access can provide services to multiple different receivers and then to different users. The principle of allocation is shown in Fig. 8.54. In this demonstration, the bandwidths of the three subchannels are all 6.25 MHz, and their center frequencies are located in 6.25, 12.5, and 18.75 MHz. The frequency spectra of a combination of different subcarriers are shown in Fig. 8.57, and the BER curve is depicted in Fig. 8.58.

The interference between the different subcarriers will deteriorate the system, so an optimal power ratio exists. For example, a combination of 6.25 and 12.5 MHz is

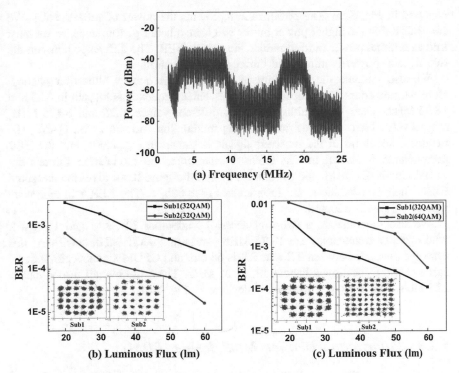

(a) Frequency (MHz)

(b) Luminous Flux (lm)

(c) Luminous Flux (lm)

Fig. 8.60 **a** Spectra of sub1 and sub3, **b** BER for sub1(32-QAM) and sub2(32-QAM), **c** BER for sub1(32-QAM) and sub2(64-QAM)

Fig. 8.61 BER performance of different modulation orders at the same bandwidth

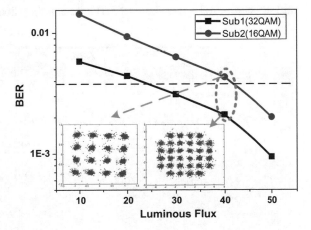

analyzed in Fig. 8.59. The power ratio represents the power of sub1 to sub3. We can realize that the higher power owner will have a better performance, so we must find an optimal power ratio to reduce the overall BER. The difference between the two signals' power is introduced during pre-equalization.

We also validate the feasibility of the flexible frequency allocation method. Here, we just adopt two subcarriers, whose center frequency is located in 6.25 and 18.75 MHz. Their bandwidths are set in different values, 6.25 and 3.125 MHz, respectively. First, they are in the same modulation formats as 32-QAM. The measured spectrum of the received signals is shown in Fig. 8.60a, and the BER performance is shown in Fig. 8.60b. Furthermore, the modulation formats are varied. One is 32-QAM, and the other is 64-QAM. Figure 8.60c gives the measured BER. Their constellations are inserted in Fig. 8.60b, c. The BER will aggravate when 64-QAM is adopted.

We also demonstrate a two-user access by adopting 32-QAM and 16-QAM, whose center frequencies are 18.75 MHz (sub1) and 33.75 MHz (sub2). At this time, the distance between TX and RX is 66 cm, and OFDM signals consist of 64 carriers, which occupy a bandwidth of 25 MHz. Thus, the overall throughput is 225 Mb/s. The BER performance is depicted in Fig. 8.61.

8.3.3 Polarization Division Multiplexing (PDM)

We propose and experimentally demonstrate a VLC system based on polarization division multiplexing (PDM) [14]. The polarization characteristic of visible light brings another degree of freedom, which can multiply the transmission capacity, compared to the aforementioned works [14]. Two orthogonal groups of polarizers and incoherent red–green–blue (RGB) light-emitting diodes (LEDs) are implemented

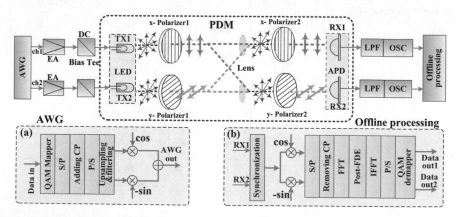

Fig. 8.62 Block diagram and experimental setup of the proposed PDM Nyquist SC-FDE VLC system (AWG: arbitrary waveform generator, P/S: parallel to serial, EA: electrical amplifier, LPF: low-pass filter, OSC: real-time oscilloscope, CP: cyclic prefix)

to realize PDM. Due to the limited experiment condition, we just utilize the red LED chip of RGB-LED. Additionally, in this demonstration, spectral efficient 16-ary quadrature amplitude modulation (16-QAM) Nyquist single-carrier frequency domain equalization (SC-FDE) is employed to obtain a high spectral efficiency (SE).

Figure 8.62 shows a block diagram of this proposed PDM Nyquist SC-FDE VLC system. As the lights emitting from incoherent LEDs are natural lights, all polarization directions are included, and they can be decomposed as two orthogonal bases of x-polarization and y-polarization, respectively. A linear x-polarizer that can only allow components in x-polarization direction passing through is implemented at transmitter1 (TX); meanwhile, a linear y-polarizer that can only allow components in y-polarization direction passing through is employed at TX2 as shown in the red box in Fig. 8.62. After passing through x-polarizer1 and y-polarizer1, we can obtain linearly polarized light, but they will be mixed up after free-space transmission. At the receiver (RX), two corresponding polarizers should be implemented to filter out the unwanted polarized lights, thus obtaining the transmitting signals.

Assuming the offset angles between x-polarizer1 and x-polarizer2, x-polarizer1 and y-polarizer2, y-polarizer1 and x-polarizer2, y-polarizer1 and y-polarizer2 are $\alpha_{11}, \alpha_{12}, \alpha_{21}$ and α_{22}, respectively. According to the Malus Law, the received optical intensity of each avalanche photodiode (APD) can be expressed as:

$$\begin{pmatrix} Y_1 \\ Y_2 \end{pmatrix} = H \begin{pmatrix} I_1 \\ I_2 \end{pmatrix} + N$$
$$= \frac{1}{2} \begin{pmatrix} \cos^2 \alpha_{11} & \cos^2 \alpha_{12} \\ \cos^2 \alpha_{21} & \cos^2 \alpha_{22} \end{pmatrix} \begin{pmatrix} I_1 \\ I_2 \end{pmatrix} + N \qquad (8.22)$$

Of which, Y_1 and Y_2 represent the received optical intensity of RX1 and RX2; meanwhile, I_1 and I_2 represent the emitted optical intensity from TX1 and TX2. H and N denote the channel matrix and noise, respectively. In this theoretical model, we just consider the distortions induced by PDM. The transmitted signals can be recovered once the channel matrix satisfied the following equation:

$$\cos^2 \alpha_{11} \cos^2 \alpha_{22} \neq \cos^2 \alpha_{12} \cos^2 \alpha_{21} \qquad (8.23)$$

And the received signals can be recovered by zero-forcing (ZF) algorithms neglecting the noise:

$$\begin{pmatrix} Y_1 \\ Y_2 \end{pmatrix} = H^{-1} \begin{pmatrix} I_1 \\ I_2 \end{pmatrix} \qquad (8.24)$$

(a)   **(b)**

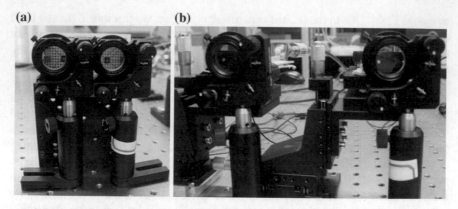

Fig. 8.63 Photograph of PDM platform **a** transmitters and **b** receivers

In this section as a proof of concept, the offset angle is set at $\alpha_{11} = \alpha_{22} = 0°$ and $\alpha_{12} = \alpha_{21} = 90°$. Therefore, the channel matrix can be simplified as a diagonal matrix, so the received optical intensity can be denoted as:

$$\begin{pmatrix} Y_1 \\ Y_2 \end{pmatrix} = \begin{pmatrix} 1/2 & 0 \\ 0 & 1/2 \end{pmatrix} \begin{pmatrix} I_1 \\ I_2 \end{pmatrix} + N \tag{8.25}$$

Neglecting the noise term N, the only difference between transmitted signals and received signals is a constant, so no extra de-multiplexing algorithm is needed.

The experimental setup is also shown in Fig. 8.62. The random binary data is generated in MATLAB and would be firstly split into two parallel streams, one for each transmitter (Tx) channel. In each channel, the bit stream is mapped into 16-ary quadrature amplitude modulation (16-QAM) format and then the training sequences (TSs) are inserted into the signals. After adding cyclic prefix (CP), up-sampling by

Fig. 8.64 BER performance versus different bias voltages

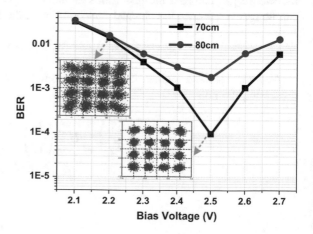

Fig. 8.65 Measured
amplitude of channel matrix
at frequency domain of
a H11; **b** H12; **c** H21; **d** H22

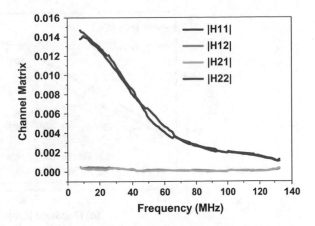

a factor of 10, and filtering by a rectangular filter, up-conversion is accomplished by
multiplying the real part and an imaginary part of signals with –sine function and
cosine function, respectively, and then added. The SC-FDE waveform is then
loaded into an arbitrary waveform generator (Tektronix 7122C). The output of this
generator is amplified by electrical amplifier (EA) (Minicircuits, 25-dB gain) and
combined with direct current (DC) bias via bias tee, and the resulting waveform is
then applied to red chip of RGB-LED acting as an optical transmitter.

Passing through x-/y-polarizer1, free space, lens (50 mm diameter, 18 mm focus
length), and x-/y-polarizer2, the signals are detected by two APDs. Then, the
received signals from each of the two RXs are routed to a high-speed oscilloscope
(Lecroy) and are acquired for further digital signal processing (DSP). After synchronization, down-converting to baseband and removing CP, the received data
streams are processed at frequency domain. The final streams are then passed
through the QAM decoder to recover the original binary stream.

The photographs of the experimental setup are depicted in Fig. 8.63. Two
commercial available RGB-LEDs (Cree, PLCC) are used as the TXs, and two
APDs (Hamamatsu, 0.5 A/W sensitivity at 800 nm) are used as the receivers RXs.
The FFT size is 128, and the CP length is 1/16. The sampling rate of AWG and
OSC is 1.25 and 2.5 GS/s, respectively. Thus, the available bandwidth is 125 MHz
ranging from 7.8125 to 132.8125 MHz. The distance between two transmitters is
5 cm.

In VLC system, LED shows significant nonlinearity due to the nonlinear characteristics of current–voltage curve and output power–current curve. Additionally,
the LED has a threshold value of about 2 V in this demonstration. The peak-to-peak
voltage after electrical amplifier is about 900 mV. When the lowest signal is below
threshold value, clipping will be introduced; meanwhile, it will work at its saturation area when the highest signal is too large. So in order to render the LED work
at the quasi-linear region, the BER performances versus different bias voltages at 70
and 80 cm are measured as shown in Fig. 8.64.

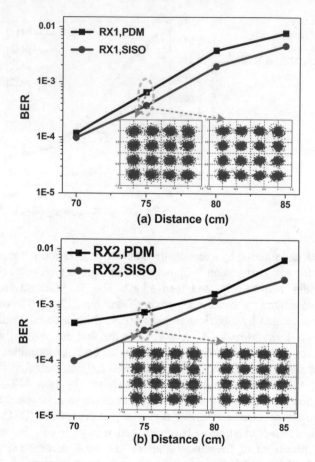

Fig. 8.66 Measured BER performance vs. different transmission distance in the case of only one polarization transmission and both polarization transmission **a** RX1 and **b** RX2

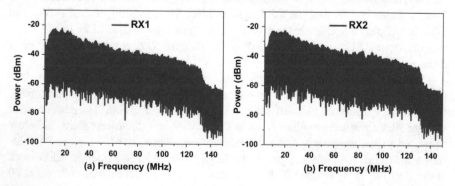

Fig. 8.67 Measured electrical spectra of **a** RX1 and **b** RX2 in the case of both polarizations transmission

Fig. 8.68 Measured BER performance versus different transmission distance with and without polarizers

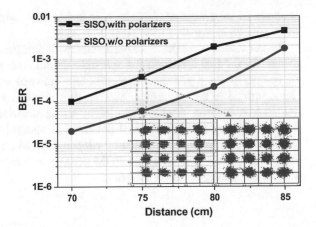

The bias voltage is varied from 2.1 to 2.7 V with a step of 0.1 V. From the measured results at both distances, the optimal working point is at 2.5 V. The constellations of 2.2 and 2.5 V at 70 cm are also inserted in Fig. 8.64. It can be easily seen that the constellation of 2.5 V is much clearer than that of 2.2 V.

The channel matrix of this system is measured via the time-multiplexed training sequences mentioned in [8]. The experimental results are shown in Fig. 8.65. The channel matrix elements $H_{ij}(i = 1, 2; j = 1, 2)$ represent the gain from jth transmitter to ith receiver. Between these four elements, H_{11} and H_{22} denote the signal gain of x–x-polarization link and y–y-polarization link, while H_{12} and H_{21} denote the cross talk between these two polarization branches. The cross talk can be neglected as it is about 100 times smaller than the signal gain.

Next, we measured the BER performance of two receivers in the case of only x-/y-polarization transmission and both polarizations transmission. The data rate of each polarization is 500 Mb/s, and thus the overall data rate is 1 Gb/s. The measured results are shown in Fig. 8.66. All BERs are under the pre-FEC limit of 3.8×10^{-3} after 80 cm transmission. And the BER performance will be slightly degraded when two polarization directions transmitted at the same time. In the case of both polarizations transmission, the electrical spectra of two RXs are also measured and illustrated in Fig. 8.67. From this figure, we can find that the power of noise floor is about -60 dBm, and gains of low-frequency part and high-frequency part are about 35 and 15 dB, respectively. Additionally, the power difference between frequencies can be pre-equalized using the method mentioned in [15], and the performance can be further enhanced.

The performance degradation after using polarizers in a single-input single-output channel is also investigated. The measured BER results as a function of distance with and without polarizers are depicted in Fig. 8.68. From this figure, we can find that after using polarizers, the transmission distance will be about 10 cm shorter at the same BER level, and in another word, the sensitivity of Rx will be reduced. But the transmission distance can become longer by employing multiple LEDs instead of only one chip in this demonstration. The constellations of these two cases are also inserted in Fig. 8.68.

In conclusion, we have experimentally demonstrated 1 Gb/s PDM in VLC system for the first time by using spectral efficient Nyquist SC-FDE and two groups of co-orthogonal polarizers. The cross talk between two orthogonal polarizers is demonstrated to be very small, and the data rate can be doubled at a sacrifice of 10 cm transmission distance shorter by using polarizers. Meanwhile, the PDM scheme utilizing two mismatched polarizers is also theoretically investigated. The data rate can be further improved by adopting WDM and a larger bandwidth APD. Additionally, a 3×3 or 4×4 PDM scheme by employing more polarizers can also be our future research direction.

8.4 The VLC MIMO

Recently, there has been constantly gaining interest in visible light communication (VLC) motivated by the dramatic development of LED technologies and increasingly scarce spectrum resources [15–17]. Compared with other typical radio frequency (RF)- based devices,widespread use, high effectivity, high brightness and larger bandwidth make it the most promising candidate for simultaneous illumination and communications especially in some specific areas like hospital, aircraft, underwater, and high-security-requirement environment. However, the relatively low modulation bandwidth is the main technical challenge in VLC system.

Many research efforts have been dedicated to overcome this limitation such as digital signal processing [15–17], high-order modulation [18], and equalizations [19]. But most of these researches are focused on single-input single-output (SISO) systems, and there have been few reports on parallel transmissions which can offer a linear capacity gain with a number of channels in an ideal cross talk-free system.

Multiple-input multiple-output (MIMO) has been widely used in radio communication and multimode fiber communication [20]. In VLC, the availability of a large number of LEDs in a single room makes MIMO a promising candidate for achieving high data rate. In [21], a comparison of nonimaging MIMO and imaging MIMO is made. The channel matrix of imaging MIMO is always full rank and diagonal matrix. But it needs precise alignment to make each LED image onto a dedicated detector. As for nonimaging MIMO, the precise alignment is not needed, and the tolerance to the misalignment becomes larger, but at the symmetry position, the channel matrix becomes less well conditioned [36]. In [22], the characterization of a diffuse-transmission MIMO is described. These examples of parallel free-space transmission are all discussed via simulation. Reference [23] experimentally demonstrates a 4×9 MIMO-OFDM VLC system with the data rate of each LED at 250 Mb/s, but the MIMO processing algorithm complexity is very high due to the large number of receivers.

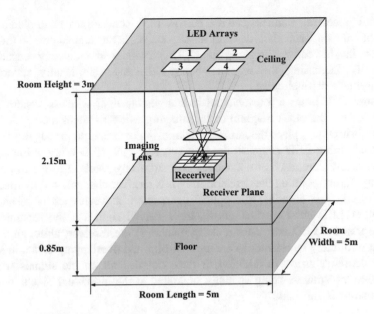

Fig. 8.69 A visible light imaging indoor wireless system model

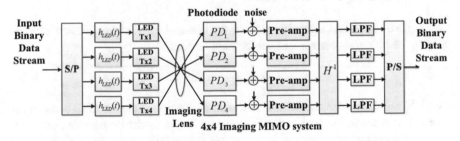

Fig. 8.70 Block diagram of the imaging MIMO system model

8.4.1 The Imaging MIMO

The use of a nonimaging optical MIMO system design is considered to be impractical, so researchers have proposed an imaging optical MIMO diversity system. The nonimaging optical MIMO system using nonimaging elements requires a separate optical concentrator for each receiving element, which may be excessively bulky and costly. The implementation of an imaging optical MIMO system using imaging optics offers two advantages over a nonimaging system. First of all, all of the photodetectors share a common concentrator, which reduces the size and cost. Secondly, all of the photodetectors can be laid out in a single array, facilitating the use of a large number of receiving elements or pixels. Since the individual receiving elements can use narrow-FOV concentrators and obtain a very high optical gain, each photodetector in the receiver is much smaller than the single detector in the conventional receiver. This reduces the detector capacitance, potentially increasing the receiver bandwidth, and significantly reduces the preamplifier's thermal noise.

Figure 8.69 shows a four-channel indoor visible light wireless system model. Figure 8.70 is the block diagram of the imaging MIMO system model. An image receiver is used to replace the nonimaging one. As before, light propagates from the four transmitting LED arrays to the receiver, and each LED array is imaged in a detector array. Here, each of the images may strike any pixels or group of pixels on the array. Every point on the detector array is a receiver channel, and by measuring the H matrix that describes the optical connection between each pixel and each transmitter LED array, it then allows the received signals to be separated. We assume that the LED arrays form sharp images at the receiver, although the processing will likely be tolerant to a degree of blur and overlapping of the images of each transmitter array. In addition, it does not depend on the signals from the individual transmitters being spatially separated at the individual pixels, only that the H matrix is full rank.

8.4.1.1 The Imaging MIMO Model

Imaging MIMO needs a modification of the channel matrix H_{image}. Each element consists of two components and can be expressed as:

$$h_{\text{image},ij} = a_{ij}h_i' \tag{8.26}$$

where h_i' represents the normalized power in the image of the ith LED array at the aperture of the imaging lens, when the receiver is at a particular position, and a_{ij} is the parameter that measures how much of this power falls on the jth receiver. h_i' can be expressed as:

Fig. 8.71 Schematic diagram of the LED arrays' image position on the receiver

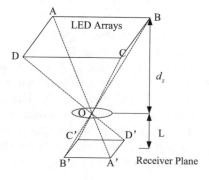

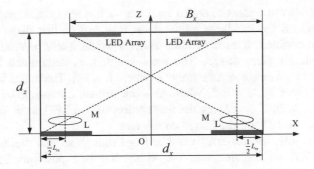

Fig. 8.72 Geometry system with the receiver at the corner of the room

$$
h'_i = \begin{cases} \sum_{k=1}^{K} \frac{A'_{rx}}{d'^2_{ik}} R_O\left(\phi'_{ik}\right) \cos\left(\varphi'_{ik}\right), & 0 \le \varphi'_{ik} \le \varphi'_c \\ 0, & x > \varphi'_c \end{cases}
\tag{8.27}
$$

where A'_{rx} is the imaging receiver collection area, d'_{ik} is the distance between the center of the receiver collection lends and the kth LED in transmitter i, ϕ'_{ik} is the emission angle, φ'_{ik} is the incidence angle of the light at the receiver, and φ'_c is the receiver FOV. An image of the ith transmitter forms in the receiver array, as shown in Fig. 8.70. Here, a_{ij} represents the proportion of the image area that falls in the jth detector pixel of the array:

$$
a_{ij} = \frac{A_{is(s=j)}}{\sum_{s=1}^{s=N_R} A_{is}}
\tag{8.28}
$$

where A_{is} represents the area of the image of the ith transmitter on the sth detector pixel in the array. In addition, the imaging position size is determined by geometric light.

Figure 8.71 is the imaging receiver's geometric position within a room. The LED arrays are imaged in the plane of the receiver by the imaging prism. In

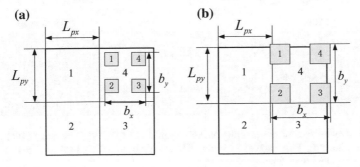

Fig. 8.73 Image of the detector arrays **a** $L_{px} > b_x$ and $L_{py} >$ by **b** $L_{px} < b_x$ and $L_{py} <$ by

this analysis, we use a paraxial optics approach, so that the system magnification is not dependent on the incidence angle of the rays and so that there is no image distortion. In addition, a more detailed analysis would likely be required for any practical optical system design. The focal length L is determined by both the diameter D and the image lens f-number $f_\#$, where $L = Df_\#$. The magnification of the system M is given by $M = d_z/L$, where dz is the vertical distance from the receiver to the ceiling. A, B, C, and D are the four corners of one LED array in the ceiling, and A', B', C' and D' are the image coordinates.

The size of the detector array must be large enough so that the images of the transmitters fall on the detectors for all the receiver positions in the room. Figure 8.72 shows the worst-case scenario when the receiver is in the corner of the room. So, L_{rx} is:

$$L_{rx} \approx \frac{2B_x}{M} \tag{8.29}$$

Here, the dimensions are as shown in the diagram and $L_{rx} \leq B_x$.
Similarly,

$$L_{ry} \approx \frac{2B_y}{M} \tag{8.30}$$

Figure 8.73 shows the schematic diagram of the detector array when the receiver is located in the center of the room. If the size of the detector pixel is $L_{px} \times L_{py}$, and $L_{px} > b_x$, $L_{py} > b_y$, as shown in Fig. 8.73a, all of the received signals will fall on a detector and the signal will not be restored. However, if each signal is received by the detector, as shown in Fig. 8.73b, the signal can be restored. Thus, a large number of transmitters and receivers using similar techniques can estimate the

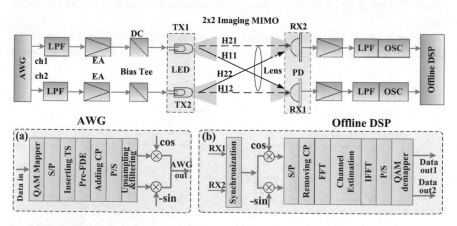

Fig. 8.74 Architecture of the imaging MIMO Nyquist SC-FDE VLC system (AWG: arbitrary waveform generator, P/S: parallel to serial, EA: electrical amplifier, LPF: low-pass filter, DC: direct current, OSC: real-time oscilloscope)

minimum pixel size. It needs to note that in order to ensure a full rank matrix H, the number of receivers (detectors) should always be more than the number of emitters. This can be used to estimate the transmitted data by a pseudo-inverse transform.

The receiver array receives ambient light from various parts of the room, which produces an unwanted photocurrent. Therefore, it is necessary to set a threshold for each detector pixel. If the received photocurrent is less than this value, the value will be set to 0 in the following calculation. Without this threshold value, the optical power will be received in the region $L_{rx}L_{ry}$, but this will limit the region where the signal and noise are coexistent, so the signal then exceeds the threshold. It can be expressed by $\sum_{i=1}^{i=N_T} \sum_{s=1}^{s=N_R} A_{is}$; therefore, this ratio can approximately estimate the received noise power.

8.4.1.2   The 2 × 2 Imaging MIMO VLC Experiment

We propose and experimentally demonstrate a gigabit 2 × 2 imaging MIMO VLC system employing Nyquist single-carrier (SC) modulation [24, 35], which is with the same spectral efficiency (SE) but low PAPR compared with OFDM. Spectral efficient 64/32-ary quadrature amplitude modulation (64-/32-QAM), together with pre- and post-equalizations, is adopted to increase the transmission data rate. We validate this MIMO scheme on three different colors in a single red–green–blue (RGB) LED. The achieved data rate of red, green, and blue color is 1.5, 1.25, and 1.25-Gb/s, respectively. The measured bit error rates (BERs) for all wavelength channels and two receivers are all below the 7% pre-forward error correction (pre-FEC) threshold of 3.8×10^{-3} after 75 cm free-air transmission.

The architecture and principle of the imaging MIMO VLC system are shown in Fig. 8.74. In this demonstration, two commercial available RGB-LEDs (Cree, red: 620 nm; green: 520 nm; blue: 470 nm) generating a luminous flux of about 61 m used as the transmitters (TXs) and two avalanche photodiode (Hamamatsu APD, 0.5 A/W sensitivity at 800 nm and gain = 1) used as the receivers (RXs) are adopted. The 3 dB bandwidths of the red LED chip and APD are 10 and 100 MHz, respectively. An imaging lens with the diameter of 76 mm and focus length of 100 mm is implemented at the front of RXs, to render each of LEDs precisely imaged onto the individual APD. The concept of Nyquist SC-FDE [25] is very similar to that of OFDM with the difference that, in SC-FDE, the inverse fast Fourier transform (IFFT) block is moved from the transmitter to the receiver. The binary data would be firstly mapped into 64-/32-QAM format, and then the training sequences (TSs) are inserted into the signals. After making pre-equalization in frequency domain and up-sampling, cyclic prefix (CP) is added and low-pass filters are used to remove out-of-band radiation.

Subsequently, the signals are amplified by electrical amplifier (EA) (Minicircuits, 25-dB gain), combined with direct current (DC)-bias via bias tee, and then applied to these three different color chips. Passing through free-space transmission, lens, APD, the signals are recorded by a commercial high-speed

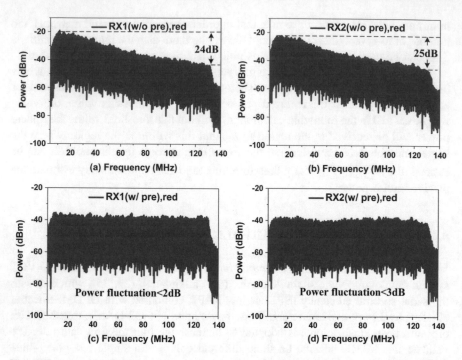

Fig. 8.75 Measured electric spectra of red LED chip at **a** RX1, w/o pre-equalization; **b** RX2, w/o pre-equalization; **c** RX1, w/pre-equalization; **d** w/pre-equalization

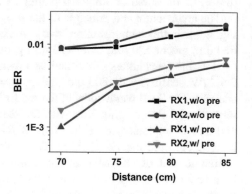

Fig. 8.76 Measured BER performance versus different transmission distances of two receivers in the case of w/and w/o pre-equalization of red LED chips

digital oscilloscope (Lecroy) and sent for off-line processing. At the receiver, after synchronization, resampling, and removing CP, post-frequency domain equalization is implemented via zero-forcing (ZF) algorithms.

The experimental setups are depicted in Fig. 8.74. The SC-FDE signals are generated by Tektronix AWG 7122C with the maximum sampling rates of 24 GS/s and bandwidths of 6 GHz and detected by a commercial high-speed real-time oscilloscope. The FFT size is 128, and the CP length is set at 1/16 of FFT size. The

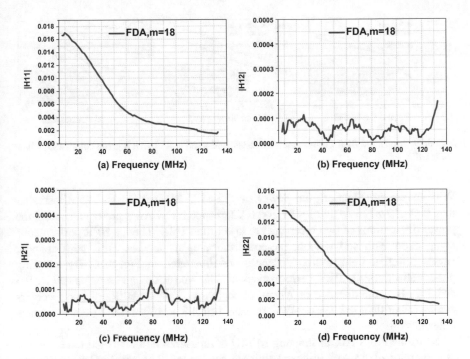

Fig. 8.77 Amplitude frequency response of **a** H11; **b** H12; **c** H21; **d** H22

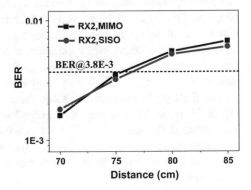

Fig. 8.78 Measured BER performance versus distance of receiver2 in the case of MIMO and SISO link of red LED chip

up-sampling factor is 10, and the sample rates of AWG and OSC are set to 1.25 and 2.5-GS/s, respectively. A square function with roll factor of 0 is used as the filter at the TX and RX. Thus, the valid occupied bandwidth of signals is 125 MHz ranging from 7.8125 to 132.8125 MHz. The voltages of bias tee and amplitudes of signals are finely adjusted to render the whole system work at the quasi-linear region of LED.

As the channel response of LED is fast attenuated as the increase of frequency, so pre-equalization is needed to improve the overall bandwidth. Figure 8.75 shows the measured electrical spectra of red LEDs before and after pre-equalization.

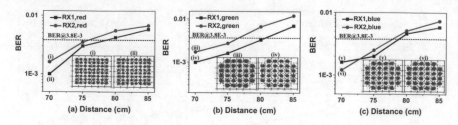

Fig. 8.79 Measured BER performance vs. distance of **a** red color LED; **b** green color LED; **c** blue color LED

Without pre-equalization, the power of 132.8125 MHz is 24 and 25 dB smaller than that of 7.8125 MHz of RX1 and RX2, respectively. After pre-equalization, the power fluctuations of two receivers are lower than 1 dB. Then, a BER performance comparison with and without pre-equalization is made. The BER performances versus different transmission distances are illustrated in Fig. 8.76. By adopting pre-equalization, the BER performances of RX1 and RX2 can be enhanced by 9.4 and 7.5 dB at the distance of 70 cm, respectively. And the BERs of two receivers can be lower than the 7% pre-FEC threshold of 3.8×10^{-3} after 75 cm free-space transmission.

Next, the cross talk of imaging MIMO is measured and analyzed based on the aforementioned time-multiplexed training sequences. The amplitudes of channel elements at frequency domain are depicted in Fig. 8.77. It should be noted that in this measurement, frequency domain averaging (FDA) proposed in [26, 27] is adopted. The index m listed in Fig. 8.77 represents the averaging window size, which should be optimized. From this figure, the values of H12 and H21 can be regarded as the additive white Gaussian noise (AWGN), which is about 100 times smaller than H11 and H22 at low-frequency component. So, we can make a conclusion that the cross talk is very limited as we analyzed in Sect. 8.2.

The BER performances of RX2 in the case of MIMO and SISO are depicted in Fig. 8.78. In SISO link, only transmitter2 works, while in MIMO links, TX1 and TX2 work at the same time. But we can find the RX2 has almost the same performance under these two conditions, which means that the degradation induced by the cross talk is very small, which can be neglected in the signal processing procedure.

At last, the pre-equalized Nyquist SC-FDE signals with bandwidth of 125 MHz are modulated on three different color LED chips. The data rates of red, green, and blue chips are 1.5, 1.25, and 1.25 Gb/s with the modulation formats of 64-QAM, 32-QAM, and 32-QAM, respectively. The BER performances versus different transmission distances ranging from 70 to 85 cm varied with a step of 5 cm are measured and depicted in Fig. 8.79a–c, respectively. Both of the BER performances of two receivers are under the pre-FEC limit of 3.8×10^{-3} after 75 cm indoor delivery.

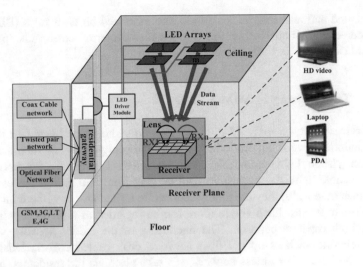

Fig. 8.80 Indoor network using nonimaging MIMO VLC

8.4.2   The Nonimaging MIMO

In this section, a 2×2 nonimaging MIMO 4-ary quadrature amplitude modulation (4-QAM) VLC system based on Nyquist single carrier is demonstrated [28]. This N-SC-FDE scheme has the similarity of spectral efficiency performance to the multicarrier modulation scheme, i.e., OFDM, and with a reduced calculation complexity compared with traditional SC scheme based on time domain equalization [28]. The de-multiplexing and post-equalization are simultaneously realized in the same step at frequency domain based on time-multiplexed training symbols. Frequency domain averaging (FDA) and time domain averaging (TDA) are also

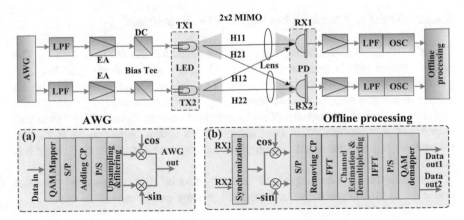

Fig. 8.81 Architecture for the proposed 2×2 nonimaging MIMO system, inset **a** detailed process of transmitter and inset **b** detailed process of off-line processing

implemented and analyzed in this paper. The measured bit error rates (BERs) for two receivers are both below the 7% pre-forward error correction (pre-FEC) threshold of 3.8×10^{-3} after 40 cm free-air transmission.

8.4.2.1 The Nonimaging MIMO Model

Multiservices with a variety of different standards should be hosted in the converged indoor network. Figure 8.80 illustrates the configuration of such a network based on visible light communication, and different terminals are hooked up through visible light. As the visible light ranging from 380 to 780 nm cannot be able to penetrate walls, they are totally confined to a single room without interfering from adjacent rooms. Each single room can be regarded as a picocell, which can provide high capacity per user. The interfacing of the indoor network with the access network, such as optical fiber network, coax cable network, twisted pair network, and 3G, 4G wireless network, can take place via the residential gateway (RG) proposed in [29], which contains media converters and possibly additional intelligence for signal conversion, local data storage, etc.

In order to maintain the typical illumination level, there would be of course many LEDs in a real room, which provide the natural setup for MIMO transmission. There are two types of MIMO system: imaging MIMO and nonimaging MIMO. The former one requires each LED array imaging onto a detector array, while the latter one only needs nonimaging concentrators before each of the receivers. In this section, the latter type is discussed and experimentally demonstrated as proof of concept. As shown in Fig. 8.80, different services passing through RG can be modulated on different LEDs via LED driver module, transmitted in the free-space environment, and detected by different receivers. After signal recovery, they can be sent to different terminals.

8.4.2.2 The 2 × 2 Nonimaging MIMO VLC Experiment

Figure 8.81 shows a block diagram of this proposed MIMO SC-FDE system. The random binary data is generated in MATLAB and would be firstly split into two parallel streams, one for each transmitter (Tx) channel. In each channel, the bit stream is mapped into M-ary quadrature amplitude modulation (M-QAM) format and then the training sequences (TSs) are inserted into the signals. The TSs in this experiment will be detailed in next section. After adding cyclic prefix (CP), and up-sampling by a factor of 10, filtering by a rectangular filter with roll factor of 0 is employed.

Assuming the obtained signals at this procedure be $X_0(f)$ at frequency domain and $X_0(t)$ at time domain, the up-converted process can be expressed as:

$$X(t) = X_{\text{real}}(t) \times \cos(2\pi f_n t) - X_{\text{imag}}(t) \times \sin(2\pi f_n t) \qquad (8.31)$$

where $X_{\text{real}}(t)$ and $X_{\text{imag}}(t)$ are the real component and imaginary component, respectively. f_n denotes the center frequency of subcarrier. $X(t)$ represents the up-converted signal in time domain, and its frequency forms can be written as

$$
\begin{aligned}
X(f) &= X_{\text{real}}(f)^* \pi(\delta(w + w_n) + \delta(w - w_n)) \\
&\quad - X_{\text{imag}}(f)^* i\pi(\delta(w + w_n) - \delta(w - w_n)) \\
&= X_0(f - f_n) + (X_0(-(f + f_n)))^*
\end{aligned}
\qquad (8.32)
$$

From Eq. (8.32), it can be seen that double sideband signals can be obtained after up-conversion. The up-conversion used here can not only provide flexible frequency allocation, but also offer radio frequency (RF) for I/Q modulation. Compared to baseband system, it can avoid low-frequency noise for the signals that are up-converted to RF frequency. After up-conversion, the SC-FDE waveform is then loaded into an arbitrary waveform generator (Tektronix 7122C) with the maximum sample rate of 24 GS/s and the maximum bandwidth of 6 GHz. The output signals of AWG are combined with a low-pass filter (LPF), which is used to remove out-of-band radiation. Subsequently, the signals are amplified by electrical amplifier (EA) (Mini-circuits ZHL-6A+, 25-dB gain), and are combined with direct current (DC)-bias via bias tee.

Passing through free-space transmission and lens (50 mm diameter, 18 mm focus length), the signals are detected by two APDs. Subsequently, the received signals from each of the two RXs are routed to a high-speed oscilloscope (Lecroy) with the maximum sample rate of 40 GS/s and the maximum bandwidth of 16 GHz

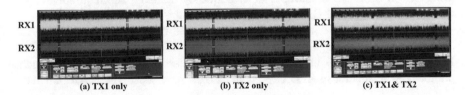

Fig. 8.82 Captured signals of two receivers in the case of **a** only TX1 transmitted, **b** only TX2 transmitted, **c** TX1 and TX2 transmitted at the same time

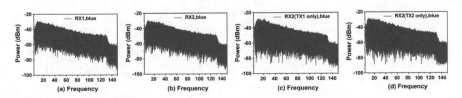

Fig. 8.83 Measured electrical spectra of **a** RX1, **b** RX2 in the case of two transmitters work at the same time and **c** RX2 (only TX1 transmitted), **d** RX2 (only TX2 transmitted)

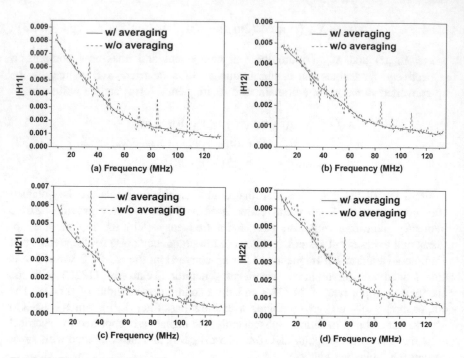

Fig. 8.84 Channel estimation of frequency matrix with and without frequency domain averaging **a** H11, **b** H12, **c** H21, **d** H22

and are acquired for further digital signal processing (DSP). After synchronization, down-converting to baseband and removing CP, the received data streams are processed in a MIMO demultiplexer at frequency domain. The final streams are then passed through the QAM decoder to recover the original binary stream.

Figure 8.81 illustrates the experimental setup of the proposed SC-FDE VLC system. In this demonstration, two commercial available blue LED chips (Cree PLCC6, blue: 470 nm) are used as the transmitters (TXs) and two avalanche photodiodes (Hamamatsu APD C5331, about 0.05 A/W sensitivity at 470 nm at the

Fig. 8.85 BER versus
different averaging window
size

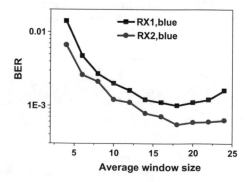

gain of 1) are used as the receivers (RXs). The dynamic range of the overall system is about −4–18 dBm. The block FFT size is 128, and the CP length is 1/16 of symbol length in this experiment. The up-sampling factor is 10, and the sample rates of AWG and OSC are 1.25 and 2.5 GS/s, respectively. Thus, the occupied electrical bandwidth of each LED chip is 125 MHz ranging from 7.8125 to 132.8125 MHz. The center frequency is located at 70.3125 MHz. The distance between two transmitters and two receivers is 5 and 10 cm, respectively, and the offset of the center positions of two sides is 2.5 cm, which will break the ill condition. The transmission distance is varied from 20 to 50 cm. The illuminance at 40 cm is about 3.5 lx which is far smaller than the indoor illuminance standard.

Fig. 8.86 BER versus distances of two receivers using 1 TS and 2 TSs

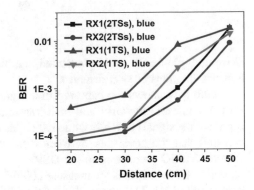

Fig. 8.87 Measured BER versus distances; inset (i–ii) shows the constellations of two different receivers

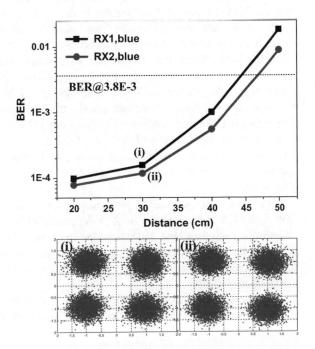

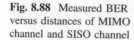

Fig. 8.88 Measured BER versus distances of MIMO channel and SISO channel

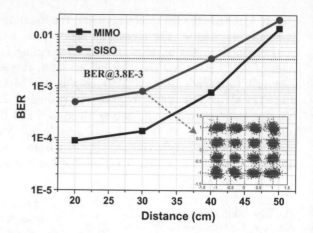

Once multiple LEDs are implemented, the illuminance can become larger, thus to enhance the transmission distance.

Figure 8.82 shows the captured time signals of RX1 and RX2 in the case that only one transmitter works and both transmitters work. The yellow colors (upper) represent the signals of RX1, while red colors (under) for RX2. From this figure, we can find that the cross talk between these two transmitters is very high, which requires MIMO processing in the DSP.

Figure 8.83 illustrates the measured electrical spectra of the two receivers in blue color LED chips. The attenuation of the highest frequency is about 30 dB larger than the lowest frequency component in this system. And the signal-to-noise ratio (SNR) in the case that both transmitters work is higher than only one transmitter work, which can be concluded by comparing Fig. 8.83a–d.

Figure 8.84 shows the amplitudes of channel matrix coefficients estimated without and with the frequency domain equalization process as a function of the frequency, under 40 cm free-space transmission of blue LED chip. The original elements of the estimated channel matrix are displayed in Fig. 8.84 as blue line. The estimated channel coefficient without the FDA exhibits high-frequency fluctuations due to the presence of the optical noise, and with FDA, the high-frequency fluctuations can be removed shown as red line. In this paper, the FDA algorithm is applied in order to make the estimated channel response much smoother (Fig. 8.84).

Figure 8.85 shows the BER performance versus different averaging window size from 4 to 24. We can find the BER performance will be improved with the increase of averaging window size, but when it is larger than 18, the BER performance will degrade. So an optimal averaging window size should be set to 18. A larger window size will reduce the frequency resolution, and a smaller one cannot remove the frequency fluctuation.

After employing FDA, the TDA is discussed in this paper. Figure 8.86 shows the BER performance versus distance with and without TDA. After using TDA with two pairs of TSs, the BER performance can be improved, but with an increase of overhead.

Next, we measured the BER performance versus different transmission distances after adopting time domain averaging and frequency domain averaging. The results are depicted in Fig. 8.87. We can find the performance of two receivers is almost the same, and the BER degrades with the increase of delivery distance. All BERs are below the 7% pre-FEC limit of 3.8×10^{-3}. The constellations of two independent tributaries in this case are also inserted in Fig. 8.87.

A comparison of the BER versus distances of a parallel MIMO transmission and single-input single-output (SISO) transmission is made. For the SISO case, the occupied bandwidth is also 125 MHz, but the modulation format is 16-QAM. Thus, the SISO transmission owns the same data rate of 500 Mb/s compared to MIMO. In this comparison, the illuminance of SISO is half of MIMO. The measured results are depicted in Fig. 8.88. From this figure, it can be seen that at the same transmission distance, the SISO case can support higher order modulation format with slightly BER penalty. This is mainly caused by the cross talk introduced by the multiple channel receivers. The constellation of the SISO transmission is also inserted in Fig. 8.88.

8.4.3 The Equal Gain Combining STBC

So far, research works on VLC are mainly focused on point-to-point transmission. For future network applications, multiple-input multiple-output (MIMO) VLC networking is considered as one of the key challenges. Space-time block coding (STBC) technique, as one of the representative multiple antenna techniques, has recently been numerically investigated and experimentally demonstrated for VLC. However, these approaches are not for MIMO-based local area network applications. For MIMO system, 2×2 STBC is widely utilized. Compared with 2×1 STBC, the MIMO algorithm achieves better performance at the cost of higher computation complexity.

In this section, we propose a novel scheme based on the equal gain combining (EGC) STBC in a high-speed 2×2 MIMO VLC network [30]. The proposed decoding scheme has the similar performance with the traditional 2×2 STBC, while the computation complexity is almost the half of the traditional one. Based on the EGC-STBC method, a 2×2 MIMO VLC network with a throughput of 1.8 Gbit/s is experimentally achieved over 1.65 m free space utilizing 16-QAM-OFDM. The results clearly validate the feasibility for future local area network applications.

Fig. 8.89 Ave CPU times
versus number of bit

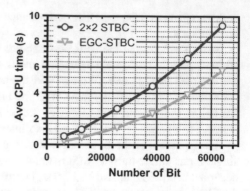

8.4.3.1   The Principle of EGC-STBC

We employ Alamouti's STBC scheme [31] to encode the data with two LED transmitters and different decoding for two receivers. The space-time block decoding can be present as follows:

$$\widehat{c} = \arg\min_{\widehat{c} \in c} \|\widetilde{r} - \rho\widehat{c}\| \tag{8.33}$$

where $\widehat{c}$ is the decoded signal and $\widetilde{r}$ is the received data. ρ is the coefficient of channel response.

For 2×2 STBC decoding, the final data is decoded by linear combination with both receivers[7]. It can be described as follows:

$$\widehat{c} = \arg\min_{\widehat{c} \in c} \|(\widetilde{r}_1 + \widetilde{r}_1) - (\rho_1 + \rho_2)\widehat{c}\|^2 \tag{8.34}$$

From the above formula derivation, it can be found that 2×2 STBC has higher computation complexity, as it needs the data of two receivers processed,

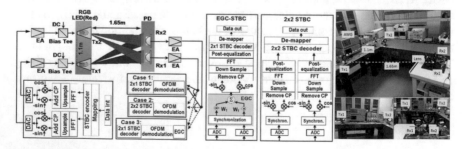

Fig. 8.90 Block diagrams and experiment picture of MIMO VLC system. Measured electrical spectra: transmitted signal of Tx1 and Tx2, received signal of Rx1

respectively. However, this computation complexity only leads to the gain of linear combination.

Therefore, we propose a modified STBC decoding which boasts less algorithm complexity. We use EGC combiner subsequent to 2×1 STBC decoding.

According to the aforesaid analysis, three different reception scenarios can be described as follows:

- For 2×1 STBC case, each receiver could calculate their own BER performance. The data is processed by Eq. (8.33) after orthogonal frequency division multiplexing (OFDM) demodulation.
- For 2×2 STBC case, the data of two receivers should be OFDM demodulated, respectively. Then, the two signals will be loaded into 2×2 STBC decoder by Eq. (8.34).
- For EGC-STBC case, two signals should be processed by EGC firstly before demodulated. The principle of EGC is to do equally weighted addition of the two detected signals after synchronization. Then, the combined signal is processed by OFDM demodulated and 2×1 STBC decoding.

Utilizing EGC-STBC, the data of two receivers will be processed by EGC before OFDM demodulation, which means it needs OFDM demodulation process only once process instead of twice. Moreover, from Eqs. (8.33) and (8.34), we can see 2×1 STBC is more simple than 2×2 STBC. Evidently, we simulate the average CPU time versus the number of bit as shown in Fig. 8.89. The measured CPU time is tested by MacBook AIR 2014 @1.4 GHz i5 including OFDM demodulation and STBC decoding. We could figure out that the CPU time of the modified STBC decoding is almost the half of the traditional one. Apparently, the proposed modified STBC scheme could be cost-effective and has less time consumption.

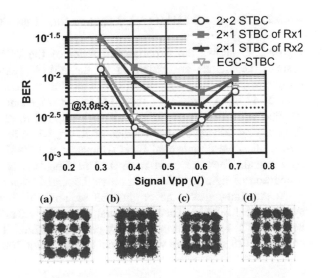

Fig. 8.91 BER versus signal drive voltage of LEDs a 2×2 STBC, b 2×1 STBC of Rx1, c 2×1 STBC of Rx2, d EGC-STBC

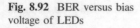

Fig. 8.92 BER versus bias voltage of LEDs

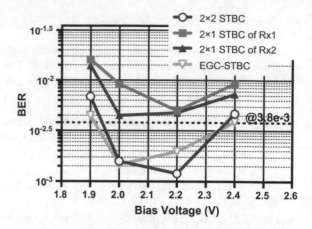

Fig. 8.93 BER versus bandwidth

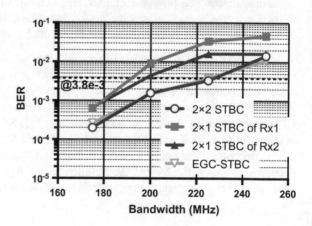

8.4.3.2 The 2 × 2 MIMO VLC System Based on EGC-STBC

Figure 8.90 shows the experimental setup of the MIMO VLC system. In this system, the signals are generated by the AWG (Tektronix AWG710) with an off-line MATLAB® program. After electrical amplifiers (EAs, Minicircuits, 25-dB gain, 50 Ω input impedance and 50 Ω output impedance), the resulting waveform coupled with direct current by bias tee is applied to red chip of two RGB-LEDs (Engin, LZ4-20MA00) at a distance of 1.1 m, while the green and blue LEDs are turned off. Two commercial PINs (Hamamatsu 10784) are used for detecting the light signal. They are apart from 0.12 m. The lens is utilized to capture a high proportion of light in order to improve the signal-to-noise ratio (SNR).

At the transmitters, the original bit sequence is firstly mapped into complex symbols of 16-QAM. Then, the signal is coded by STBC into two signal sources before OFDM modulation and soft pre-equalization. The STBC-OFDM signals consist of 128 subcarriers with channel bandwidth of 200 MHz, and the data rate of each LED is 800 Mbit/s using 16-QAM operating at the distance of 1.65 m, respectively. The whole throughput is 1.6 Gbit/s. In this experiment, up-sampling by a factor of four is employed, and the sample rate of AWG is 800 MS/s while the digital real-time oscilloscope (OSC, Agilent 54855A) is 1 GS/s. The off-line digital signal processing (DSP) is applied to demodulate the sampled signal by OSC.

In the off-line process, for EGC-STBC scheme, the two data streams of receivers are firstly processed by EGC. Then, the data needs OFDM demodulation before 2×1 STBC decoding. While for 2×2 STBC scheme, the received data of Rx1 and Rx2 is both demodulated by OFDM and then sent to STBC decoder.

In the MIMO VLC system, we measure the influence of signal driving voltage and bias voltage of LEDs. Figure 8.91 shows the measured BERs versus the signal driving voltage of AWG. We set the signal driving voltage of both LEDs the same. According to the measured results, the optimal signal driving voltage of LEDs is 0.5 Vpp. Figure 8.92a–d is constellations of 2×2 STBC, 2×1 STBC of Rx1, 2×1 STBC of Rx2, and EGC-STBC, respectively. The performances of Rx1 and Rx2 are a little different because the circuit of photo-detectors could not be equal.

Figure 8.92 shows the measured BERs versus the bias voltage of LEDs. The bias voltage for both LEDs is set the same to simulate a practical indoor lighting environment. The optimal bias voltage of LEDs is 2V. While increasing the voltage, the luminance is enhanced, but the received PINs would be into saturation. From these two figures of BER performance, we could find that the performance of EGC-STBC and 2×2 STBC is much better than 2×1 STBC. Besides, these two approaches have a similar BER performance in MIMO VLC system.

Figure 8.93 shows the BER versus the modulation bandwidth. The modulation bandwidth of 225 MHz, which means a throughput of 1.8 Gbit/s, is successfully achieved with BER under 3.8×10^{-3}. Among the whole experiment, our proposed EGC-STBC decoding works as well as the traditional 2×2 STBC decoding in the 2×2 MIMO VLC system. However, the proposed method has much less computation complexity than the traditional one. The disparity of the operating CPU time in the actual experiment is the same as the simulation result shown in Fig. 8.89.

8.5 The VLC Network

In order to realize the practical application of visible light communication, one of the important problems is the visible light communication network. Since visible light communication is regard to as "the last few meters" wireless access technology, how to merge with the backbone of the existing networks, how to implement in a large office environment, hundreds of users' high-speed wireless access is a key problem to be solved. Up to now, there are three ways to combine VLC with other access techniques, and one of them is the integration of VLC and power line communication (PLC). Because VLC uses LED as light source, providing lighting at the same time, it is quite natural to combine VLC with PLC. It is convenient to provide electricity supply for LED through the power line, transmitting the uplink

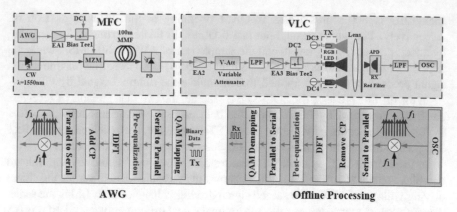

Fig. 8.94 Architecture of the integrated network of VLC and MMF

and downlink data stream at the same time. But the main problem in this way is that PLC is easily affected by the electrical environment, thus bringing the signal fluctuation. Meanwhile, modulation bandwidth of PLC is very limited, only providing data rate of hundred megabits, so it cannot meet the future demand for more than 10 Gb/s wireless access services [5]. The second way is the integration of VLC and Ethernet. It is greatly compatible with the current local area network (LAN) and easy to implement at low cost. But the main problem in this way lies in the restricted Ethernet access rate, and thus it cannot satisfy the demand of high-speed wireless transmission. The last way is the integration of VLC and the optical fiber network. Although it would increase the cost of the system in this way, it can fully utilize the advantages of large capacity and wide bandwidth in optical network, and then support a large number of users at the same time. This scheme is prone to be an important way to implement VLC in the future.

This section is about the key problems of VLC network. It will introduce the combination of VLC and multimode fiber (MMF), VLC and the passive optical network (PON), VLC and the optical fiber as the backbone architecture of multi-user access networks, and also introduce the related experiment systems and the experimental results.

8.5.1   The Integrated Network of VLC and MMF

Multimode fiber has been widely used in indoor access network due to its advantages of low cost, easy deployment, and large transmission capacity. Combining VLC system with MMF can further promote their relative advantages, thus achieving high-speed wireless access of multiple users. Figure 8.94 depicts the architecture of the integrated network of VLC and MMF.

Figure 8.94 shows the block diagram for integrated VLC and MFC system, respectively. In this scheme, the system consists of VLC system and MFC system.

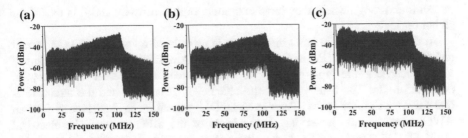

Fig. 8.95 Measured electrical spectra of **a** original signal, **b** after MFC, and **c** after APD

In the MFC system, the driving signal from arbitrary waveform generator (AWG, Tektronix AWG710) is amplified by EA1 (JDS Uniphase, Model H301) and then coupled with DC1 by Bias Tee1 (Picosecond, Model 5575A). The output of Bias Tee1 and the continuous wave (CW) light beam from a laser diode is fed into a Mach-Zehnder modulator (MZM, Lucent 2623NA) which is connected with 100 m MMF. Then, the signal from MMF is detected by a photodetector (PD).

After converting light signal to electronic signal by the PD, the resulting waveform is applied to the input of the VLC system amplified by two electrical amplifiers, EA2 and EA3 (Minicircuits, 25-dB gain). And a variable attenuator between EA2 and EA3 is used to adjust the input power of red LED for better performance of the whole system. The transmitted signal is carried by red light of RGB-LED (Cree, PLCC) radiated in the wavelength regions of 625 nm, while the green and blue LEDs are turned on to maintain white color illumination with only DC supply and both voltages are 3.3 V. A commercially available high sensitivity silicon APD (Hamamatsu APD, 0.5 A/W sensitivity at 800 nm and gain = 1) combined with a lens (76 mm diameter) in front of it is used for detecting the light signal. The lens is used to capture a high proportion of light in order to improve the signal-to-noise ratio (SNR) of the system. And the red optical filter is implemented to filter out the noise light wavelength, like green and blue light. Then, reason why we choose red light is that most of PINs and APDs are designed for infrared and

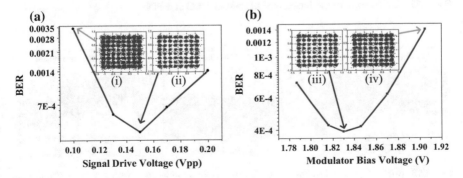

Fig. 8.96 Measured BERs versus **a** signal drive voltage and **b** bias voltage of modulator

near-infrared range where they have extremely high sensitivity and it is useful for signal retrieving.

At the transmitter, the input binary data is modulated by using 64-QAM format and then passed to OFDM encoder. Then, the QAM-OFDM signals consisting of 128 subcarriers are up-converted to the carriers with center frequency at $f = 56.25$ MHz. The bandwidth of the channel is 100 MHz, and the data rate is 600 Mbit/s using 64-QAM. The QAM-OFDM waveform is then loaded into an AWG, whose output is serving as the input of the MFC system. The electrical QAM-OFDM signals and DC bias voltage are combined via bias tee and applied to the red LEDs acting as the transmitter. Up-sampling by a factor 10 is employed, and the sampling rate of AWG is 1.00 GSa/s. At the receiver, the optical signal is detected by the APD and recorded by a real-time digital oscilloscope (OSC, Agilent 54855A) whose sampling rate is 1.00 GSa/s. A control computer is used to acquire waveform signals from it for subsequent further off-line procession to retrieve the original signal. Then, the received signals are down-converted to baseband and processed with the inverse procedure of QAM-OFDM encoder.

Figure 8.95a, b shows the measured electrical spectra of original signal and output signal after MFC in frequency domain, while (c) is the whole system of MFC and VLC at 60 cm. The spectra in (a) and (b) have the same trend rising from 6.25 to 106.25 MHz because of the pre-equalization used to compensate in high frequency. Compared with (a), (b) has small attenuation in high frequency, while the spectrum in (c) is flat for the high-frequency fading in the VLC system. So, the dominant effective factor on the whole system depends on the VLC part.

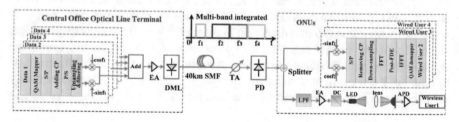

Fig. 8.97 Architecture of the integrated network of VLC and PON

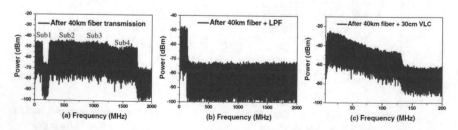

Fig. 8.98 Subcarrier electrical spectra **a** after transmission over 40-km fiber, **b** after 40-km fiber and LPF, **c** after 40-km fiber and 30 cm VLC

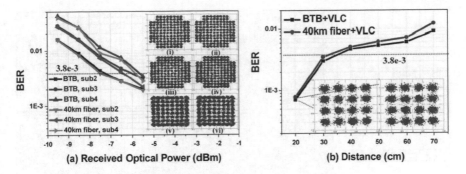

Fig. 8.99 BER performance with **a** received optical power and **b** free-space distance

In the whole system demonstration, we measured the BERs versus the signal driving voltage of AWG, bias voltage of modulator, and the distance between LED and APD is 60 cm. For driving signal voltage of AWG demonstration, the bias voltages of modulator and red LED, received power of PD, and input power of red LED are 1.83, 2.38 V, −3, and 18 dBm, respectively, and the signal driving peak-to-peak voltage (Vpp) is set from 0.10 Vpp to 0.20 Vpp. The BER versus signal drive voltage is shown in Fig. 8.96a, and the optimal voltage is 0.15 Vpp. Then, signal driving voltage is set to 0.15 Vpp and the modulator bias voltage of Bias Tee1 changes from 1.787 to 1.905 V while other parameters are unchanged. The result is shown in Fig. 8.96b, and the optimal bias voltage for modulator is 1.836 V. In the MFC, the input signal from AWG is amplified by EA1 and coupled with Bias Tee1 acting as the electrical input of MZM. The amplitude of original signal and voltage of Bias Tee1 will change the working performance of MZM, and 0.15 Vpp of driving signal and 1.836 V of Bias Tee1 are the best parameters for MFC system.

8.5.2 The Integrated Network of VLC and PON

Passive optical network (PON) has become the backbone of the high-speed wired access network. In order to meet the future demand for a variety of high-speed access, for the first time, we propose the integration of VLC and PON. With this scheme, it can simultaneously provide high-speed cable access and wireless access services for multiple users. The systematic diagram of this scheme is depicted in Fig. 8.97.

It achieves the multiple users access with subcarrier multiplexing (SCM) and the integrated network of VLC and MMF. In the setup, three subcarriers with 500 MHz bandwidth are used as wired access of PON and the subcarrier with 125 MHz bandwidth on the low-frequency part is used to demonstrate the transmission of VLC. The four subcarriers are emerged through the adder and then drive the

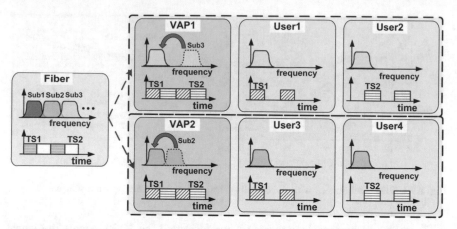

Fig. 8.100 Schematic diagram of hybrid access network protocol

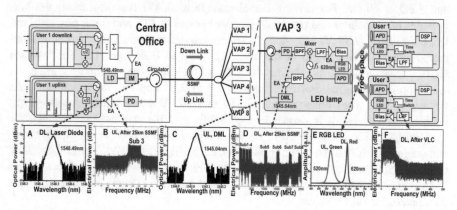

Fig. 8.101 Schematic diagram of VLC access network

directly modulated laser (DML). After transmission over 40-km fiber, the signal is detected by photonic detector. Passing through the low-pass filter and the electrical amplifier, the low-frequency subcarrier is sent into VLC transmitter to realize the wireless transmission. The other three subcarriers are demodulated to realize the wired high-speed access. The QAM SC-FDE signal with different orders is adopted in this experiment, and it achieves 10 Gb/s PON transmission and 500 Mb/s VLC transmission.

Figure 8.98 shows the electrical spectra of subcarriers, including the spectrum after transmission over 40-km fiber from Fig. 8.98a, the spectrum after low-pass filter from Fig. 8.98b and the spectrum after VLC channel from Fig. 8.98c.

Figure 8.99a shows the BER performance with optical received power over four subcarriers. And Fig. 8.99b shows the BER performance with free-space distance for different channels.

Fig. 8.102 BER performance of eight subcarriers

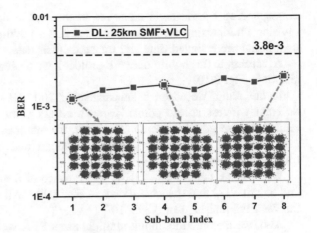

Fig. 8.103 Uplink two-user BER performance versus received optical power

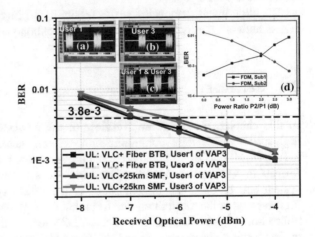

8.5.3 The High-Speed Full-Duplex VLC Access Network

In the process of promoting VLC into practical application, a critical problem is how to build a high-speed VLC network that can support the Gb/s wireless access for a large number of users at the same time. In order to achieve this goal, we put forward the star topology structure of VLC local access networks. By using optical fiber as the network framework, the data from the central station directly connects to the LEDs, each LED as a wireless access point, realizing the multi-user access.

In order to take full advantage of large capacity and large bandwidth of optical fiber at the same time to realize multi-user access for each wireless access point, we design a hybrid access network protocol, the protocol of the schematic diagram as shown in Fig. 8.100. In the protocol, we adopt frequency division multiplexing as optical fiber uplink/downlink scheme and each access point is allocated to corresponding subcarrier in advance. In the visible light access, we adopt time division multiplexing technology to realize the multi-user wireless access. Each user in the pre-assigned time slot transmits their data.

According to this hybrid network protocol, we design the VLC access system as shown in Fig. 8.101.

In the setup, we adopt 8 subcarriers of 100 MHz bandwidth allocated to 8 different wireless access points. In each access point, two users simultaneously realize high-speed VLC access. Therefore, the total capacity reaches 8 Gb/s. Each user could get access to 500 Mb/s download and 500 Mb/s upload wireless transmission service.

We measure the downlink BER performance of 8 subcarriers after 25-km fiber and 65 cm VLC free space as shown in Fig. 8.102. All subcarrier BERs are below the 7% FEC threshold of 3.8×10^{-3}.

Also, we measure the uplink multiple users BER versus received optical power as shown in Fig. 8.103.

The experiment demonstrates the feasibility of high-speed full-duplex optical access network that is based on fiber optic backbone architecture.

8.6   Summary

In this chapter, we make an overview of the advanced techniques used in the high-speed visible light communication system, including SCM, WDM, PDM, MISO, MIMO, STBC, and bidirectional transmission. Firstly, advanced modulation formats in VLC system are introduced in Sect. 8.1. Section 8.2 introduces MISO system and bidirectional transmission. In the MISO system, multiple LEDs are implemented at the transmitter, and a single PD is implemented at the receiver. The bidirectional transmission is one of the VLC's technical challenges. Time division or frequency division can be used to reduce the interference from different links.

By adopting multidimensional multiplexing techniques in Sect. 8.3, a high-speed transmission can be realized. SCM utilizes multiple subcarriers to transmit signals, and additionally, flexible frequency allocation and multiple access can be realized by adjusting the center frequency, bandwidth, and modulation order of each subchannel. WDM makes full use of the three different RGB-LED colors and can triple the capacity.

In VLC, the availability of a large number of LEDs in a single room makes MIMO a promising candidate for achieving high data rate. Section 8.4 clarifies imaging MIMO, nonimaging MIMO, and STBC VLC system. Section 8.5 briefly introduces potential VLC networks integrated with other techniques, such as MMF, PON, and full duplex.

References

1. Li, J., Huang, Z., Liu, X., Ji, Y.: Hybrid time-frequency domain equalization for LED nonlinearity mitigation in OFDM-based VLC systems. Opt. Exp. **23**(1), 611–619 (2015)
2. Wang, Y., Tao, L., Huang, X., Shi, J., Chi, N.: Enhanced performance of a high speed WDM CAP64 VLC system employing Volterra series based nonlinear equalizer. IEEE Photonics J. **7**(3), 1–7 (2015)
3. Cossu, G., Khalid, A.M., Choudhury, P., Corsini, R., Ciaramella, E.: 3.4 Gbit/s visible optical wireless transmission based on RGB LED. Opt. Exp. **20**(26), B501–B506 (2012)
4. Tao, L., Wang, Y., Gao, Y., Lau, A.P.T., Chi, N., Lu, C.: 40 Gb/s CAP32 system with DD-LMS equalizer for short reach optical transmissions. IEEE Photonic Technol. Lett. **25** (23), 2346–2349 (2013)
5. Tao, L., Wang, Y., Gao, Y., Lau, A.P.T., Chi, N., Lu, C.: Experimental demonstration of 10 Gb/s multilevel carrier-less amplitude and phase modulation for short range optical communication systems. Opt. Exp. **21**(5), 6459–6465 (2013)
6. Huang, X., Shi, J., Li, J., Wang, Y., Wang, Y., Chi, N.: 750 Mbit/s visible light communications employing 64QAM-OFDM based on amplitude equalization circuit. In: Optical Fiber Communication Conference (OFC), Tu2G.1 (2015)
7. Shafik, R.A., Rahman, M.S., Islam, A.H.M.: On the extended relationships among EVM, BER and SNR as performance metrics. In: Proceedings of International Conference on ICECE, pp. 408–411 (2006)
8. Wang, S.W., Chen, F., Liang, L., He, S., Wang, Y., Chen, X., Lu, W.: A high-performance blue filter for a white-led-based visible light communication system. IEEE Wirel. Commun. **22**(2), 61–67 (2015)
9. Huang, X., Shi, J., Li, J., Wang, Y., Chi, N.: A Gbps VLC transmission using hardware pre-equalization circuit. IEEE Photonics Technol. Lett. **27**(18), 1915–1918 (2015)
10. Fischer, R.F., Huber, J.B.: A new loading algorithm for discrete multitone transmission. In: Proceedings of International Conference on GLOBECOM'96. Communications: The Key to Global Prosperity, vol. 1, pp. 724–728 (1996)
11. Liu, Y.F., Yeh, C.H., Chow, C.W., Liu, Y., Liu, Y.L., Tsang, H.K.: Demonstration of bi-directional LED visible light communication using TDD traffic with mitigation of reflection interference. Opt. Exp. **20**(21), 23019–23024 (2012)
12. Wang, Y. et al.: 875-Mb/s asynchronous bi-directional 64QAM-OFDM SCM-WDM transmission over RGB-LED-based visible light communication system. In: Optical Fiber Communication Conference on Optical Society of America (2013)
13. Cossu, G., Khalid, A.M., Choudhury, P., Corsini, R., Ciaramella, E.: 3.4 Gbit/s visible optical wireless transmission based on RGB LED. Opt. Exp. **20**(26), B501–B506 (2012)
14. Wang, Yuanquan, Yang, Chao, Wang, Yiguang, Chi, Nan: Gigabit polarization division multiplexing in visible light communication. Opt. Lett. **39**(7), 1823–1826 (2014)
15. Wang, Y., Wang, Y., Chi, N., Yu, J., Shang, H.: Demonstration of 575-Mb/s downlink and 225-Mb/s uplink bi-directional SCM-WDM visible light communication using RGB LED and phosphor-based LED. Opt. Exp. **21**(1), 1203–1208 (2013)
16. Wang, Y., Chi, N., Wang, Y., et al.: High-speed quasi-balanced detection OFDM in visible light communication. Opt. Exp. **21**(23), 27558–27564 (2013)
17. Li, R., Wang, Y., Tang, C. et al.: Improving performance of 750-Mb/s visible light communication system using adaptive Nyquist windowing. Chin. Opt. Lett. **11**(8) (2013)
18. Wu, F.M., Lin, C.T., Wei, C.C. et al.: 3.22-Gb/s WDM visible light communication of a single RGB LED employing carrier-less amplitude and phase modulation. In: Proceedings of OFC 2013, OTh1G.4
19. Wang, Y., Shao, Y., Shang, H., Lu, X., Wang, Y., Yu, J., Chi, N.: 875-Mb/s asynchronous bi-directional 64QAM-OFDM SCM-WDM transmission over RGB-LED-based visible light communication system. In: Proceedings of OFC 2013, OTh1G.3

20. Wang, Y., Chi, N., Li, R. et al.: Theoretical and simulation analysis of a novel multiple-input multiple-output scheme over multimode fiber links with dual restricted launch techniques. Opt. Eng. **51**(6), 065002-1–065002-7 (2012)

21. Lubin, Z., et al.: High data rate multiple input multiple output (MIMO) optical wireless communications using white led lighting. IEEE J. Sel. Areas Commun. **27**(9), 1654–1662 (2009)

22. Jivkova, S., Hristov, B.A., Kavehrad, M.: Power-efficient multi spot diffuse multiple-input-multiple-output approach to broad-band optical wireless communications. In: IEEE Transactions on Vehicular Technology, vol. 53, pp. 882–889 (2004)

23. Azhar, A.H., Tran, T.A., O'Brien, D.: A gigabit/s indoor wireless transmission using MIMO-OFDM visible-light communications. IEEE Photonics Technol. Lett. **21**(15), 1063–1065 (2009)

24. Wang, Y., Chi, N.: Indoor gigabit 2 × 2 imaging multiple-input multiple-output visible light communication. Chin. Opt. Lett. **12**(10), 100603 (2014)

25. Falconer, D., Ariyavisitakul, S.L., Benyamin-Seeyar A., Eidson, B.: Frequency domain equalization for single-carrier broadband wireless systems. IEEE Commun. Mag. **40**(4) (2002)

26. Li, F., Cao, Z., Li, X. et al.: Fiber-wireless transmission system of PDM-MIMO-OFDM at 100 GHz frequency. J. Lightw. Technol. **31**(14), 2394–2399 (2013)

27. Liu, X., Ling, X. et al.: Intra-symbol frequency-domain averaging based channel estimation for coherent optical OFDM. Opt. Exp. **16**(26), 21944–21957 (2008)

28. Wang, Y., Chi, N.: Demonstration of high-speed 2 × 2 non-imaging MIMO Nyquist single carrier visible light communication with frequency domain equalization. J. Lightwave Technol. **32**(11) (2014)

29. Koonen, A.M.J., Larrodé, M.G.: Radio-over-MMF techniques—Part II: microwave to millimeter-wave systems. J. Lightw. Technol. **26**(15), 2396–2408 (2008)

30. Shi, J. et al.: Improved performance of a high speed 2 × 2 MIMO VLC network based on EGC-STBC. In: European Conference on Optical Communication (ECOC), IEEE (2015)

31. Alamouti, S.M.: A simple transmit diversity technique for wireless communication. IEEE J. Select. Areas Commun. **16**(8), 1451–1458 (1998)

32. Wang, Y., Wang, Y., Chi, N.: Experimental verification of performance improvement for a gigabit WDM visible light communication system utilizing asymmetrically clipped optical OFDM. Photonics Res., **2**(5), 138–142, (2014)

33. Wang, Y., Wang, Y., Tao, L., Shi, J., Chi, N.: Experimental demonstration of a novel full-duplex high-speed visible light communication access networks architecture based on frequency division multiplexing. Opt. Eng. **53**(11): 116104–116104 (2014).

34. Wang, Y., Chi, N.: Asynchronous multiple access using flexible bandwidth allocation scheme in SCM-based 32/64QAM-OFDM VLC system. Photonic Netw. Commun. **27**(2), 57–64 (2014)

35. Wang, Y., Chi, N.: Indoor gigabit 2x2 imaging multiple-input multiple-output visible light communication. Chin. Opt. Lett. **12**(10):12–15 (2014)

36. Wang, Y., Zhou, Y., Gui, T., Zhong, K., Zhou, X., Wang, L., Lau, A.P.T., Lu, C., Chi, N.: Efficient MMSE-SQRD based MIMO decoder for SEFDM based 2.4-Gb/s spectrum compressed WDM VLC system. IEEE Photonics J. **8**(4): 1–9 (2016)

Chapter 9
Visible Light Communication Technology Development Trend

9.1 Surface Plasma LED

LEDs offer many advantages including threshold-less operation, high fabrication yield, high energy efficiency, fast response and reduced complexity of driver, reducing the requirement for threshold feedback control significantly, thus reducing the overall cost, form factor, and power consumption of transmitters. High efficiency and high-speed LEDs are attractive solutions in optical data transmission, which simultaneously support illumination and communication. Although the industry has manufactured blue LEDs of high emission efficiency [1–3] nowadays, that is not the case for green LEDs, the efficiency of which is still quite low due to the poor crystal quality of the grown InGaN/GaN quantum wells (QWs) [4]. Moreover, the enhancement of the emission rate is of great significance also for the development of high-speed communication technology and optical computing. Surface plasmon (SP) is an effective way to increase the carrier radiative recombination rate, which is attributed to the new energy transition channel of electron–hole pairs in LEDs created by the QW–SP coupling. Since the density of SP mode states is large, the QW–SP coupling rate is very fast, increasing the radiative recombination rate and decreasing the carrier recombination time accordingly [5–14].

Plasmonics is the key technique to control and utilize the SP generated around metal interfaces. The SP-enhanced photoluminescence (PL) intensities are due to the increasing of the radiative recombination rates [15] through which the high-efficiency and high-speed light emission is predicted even for electrically pumped light-emitting devices; e.g., efficient light emissions have been achieved by using this technique [1]. At the same time, it is necessary to study the LED device technology, including the design of lithography layout pattern, electrode graphics, and enhanced current expansion, reduce parasitic resistance and capacitance, and improve the modulation bandwidth. Figure 9.1 shows a surface plasmon LED through self-assembled Ag NPs in the vertical proximity to the multiple quantum

© Tsinghua University Press, Beijing and Springer-Verlag GmbH Germany 2018
N. Chi, *LED-Based Visible Light Communications*, Signals and Communication
Technology, https://doi.org/10.1007/978-3-662-56660-2_9

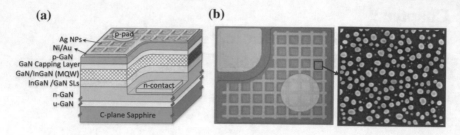

Fig. 9.1 a SP-LED device; **b** top view of grid LED with Ag NPs; the inset shows the SEM image of Ag NPs

wells (MQWs) region. Compared to the LED without SP coupling, the optical modulation bandwidth of the SP-LEDs with Ag annealed film increased by a factor of ~ 2.75, while the optical power increased by a factor of ~ 1.96 at forward current of 60 mA due to the QW–SP effective coupling. This observation highlights the importance of properly designing the diameters of Ag NPs for optimal light emission efficiency and optical modulation bandwidth enhancement. The findings can be integrated within a full LED device and therefore are promising for the development of solid-state light sources of highly improved efficiency and modulation rate for application in high-speed visible light communication.

9.2 Visual Imaging Communication

The visible light-based imaging communication system uses an liquid crystal display (LCD) or a light-emitting diode (LED) array as a transmitter for performing parallel transmission in the visible light channel after intensity, space, etc. The transmitting image information is received by imaging on the imaging device on the receiving end. Typical receivers are charge-coupled device (CCD), CMOS imagers, and photodiode (PD) arrays. After the processing of the received image, the system untunes the data that was originally sent.

When it comes to optical imaging communication, first of all, we should start with nonimaging visible light communication. The geometric shape of the non-imaging VLC MIMO (NIVLC-MIMO) system is based on the paper [16]. The system has NT LED emitters and NR receivers, the light from each LED array is received by the independent receiver, but the received strength is different. The LED array is used to illuminate the entire room. Each LED array sends parallel independent data streams. Independent data streams are acquired by MIMO signal processing techniques. The optical signal is propagated indoors and is detected by the imaging receiver. Each receiver has a nonimaging condenser. The detector array includes multiple pixel points, which are independent reception cells.

The schematic diagram of the nonimaging visible MIMO system is shown in Fig. 9.2. System input is a serial data stream that is converted to a large number of

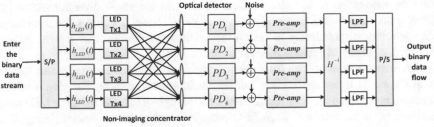

Fig. 9.2 Schematic diagram of nonimaging visible light MIMO system

parallel data streams. The number is consistent with the number of emitters. Light travels from a LED to a receiver. There are usually two forms of dissemination. Line of sight part of each LED is propagated to the receiver, and there are diffuse reflections from surface reflections in space.

When the receiver is placed in the corner of the room, compared with the direct light, diffuse part is the strongest. However, the received power is at least 7 dB lower than the weakest part of the range, so the visibility portion is considered. In this case, the channel bandwidth is affected by the relative delay intensity of the different LED video distance components and each receiver. The maximum path length between the stadia elements is in the corner of the receiver room, about 3 m, and it is approximately equal to 10-ns time delay or 100 MHz bandwidth.

A nonimaging visible MIMO system can theoretically obtain a symmetric channel matrix H, which is not full rank and is irreversible. For MIMO, the H matrix that operates must be a full rank matrix. The symmetry of the channel matrix in the nonimaging MIMO system means that the matrix in the space center and the plane coordinate axis are not satisfied with the full rank condition. When the matrix becomes unsatisfied, the error rate increases, because the matrix coefficient and the noise bias are in the same order. By rearranging the geometry of the receiver, the symmetry may be broken. We can get the right operation in all locations, but it can happen as follows that it will also produce some locations unable to break the symmetry or to produce full rank matrix. For this reason, nonimaging is considered impractical. An imaging system which can overcome these limitations will be introduced below.

Since the use of nonimaging to design a light MIMO system is considered impractical, the researchers propose using an imaging subset light MIMO system. The use of a nonimaging unit to achieve an angular set requires a separate concentrator for each receiving unit, which can be cumbersome and costly. There are two advantages using an angular diversity receiver imaging prism compared with using nonimaging ones. First, all optoelectronic detectors share a common concentrator to reduce the design size and cost. Second, all optoelectronic detectors can be placed on a single planar array, which is conducive to the use of a large number of receiving units or pixels. The receiving unit can be used with high gain narrow field of condenser, so the photoelectric detector in the angle diversity receiver is smaller than the traditional receiver in a single detector. It reduces the detector of the capacitance, potentially increases the receiver bandwidth, and greatly reduces the thermal noise of preamplifier [17].

An imaging receiver can fully decode data from each luminous element located in any position in the space under constraints in geometry space. Imaging receivers replace nonimaging devices. Same as before, the light transmits from the LED array transmitter to the receiver. Each LED array is imaging on the detector array. The image could be any of the pixels in the array or any group of pixels. Every pixel on the detector array is a receiver channel. By measuring the H matrix that describes the optical connection between each pixel and each LED array transmitter, the characteristic of the received signal can be separated. In a carefully designed optical system, the spacing of the detector array and the number of detectors can make the H matrix full rank in a wide range of conditions. The schematic diagram of the imaging optical MIMO system is shown in Fig. 9.3.

After receiving the signal, the matrix H inverse is multiplied by the receiving signal, and then the estimated value of the transmitting signal is obtained. After the estimation of the emission signal is completed, low-pass filtering and equalization are adopted for the signal. The data flow is then merged to generate the serial received data stream, which is used to get the error rate (BER) compared to the sent data stream.

The structure diagram of visible image communication system is shown in Fig. 9.4.

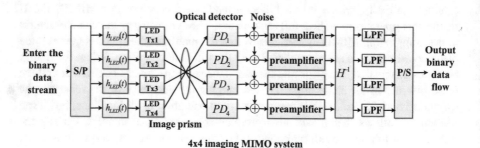

Fig. 9.3 Schematic diagram of imaging MIMO system

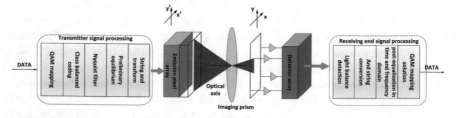

Fig. 9.4 Structure block diagram of optical imaging communication system

The optical imaging communication system mainly consists of transmitter signal preprocessing, emitter array, imaging lens, detector array, and receiving signal processing. The implementation of the system is as follows: Using the LED array or LCD as the transmitter, the electrical signal is loaded into the transmitting array after specific pretreatment and modulation. It can be launched simultaneously in the form of multichannel light signals. The receiving end of the system is an array of photoelectric detection devices. The array can receive multiple light signals simultaneously. Each light signal will be irradiated through the free-space channel to each photodetector on the receiving array. The receiving end data acquisition chip will send the collected analog signal to the signal processing chip and carry out back-end digital signal processing.

9.3   Key Issues of VLC Networking

The key problems of optical communication network are visible light source layout, visible network switching technology, and visible network access control [18].

9.3.1   Visible Light Source Layout

In practical application, the room often needs multiple sets of light source to achieve effective coverage, so the rational design of the light source layout is undoubtedly a key factor affecting the overall performance of the system.

The light intensity and power of a single LED are small. In order to realize the dual functions of indoor lighting and communication, a luminous array composed of multiple LED modules is generally adopted. Two aspects need to be considered in the layout of the light source: the layout of the LED lights in the interior of the light source array and the overall layout of the indoor LED. In the design of lighting and communication indoor light source, the rational layout of the two aspects can meet the lighting and communication needs.

For the distribution of the light source array, the total number and arrangement of LEDs needed for each light source array should be based on factors such as indoor exposure range, LED spacing, and direct distance. It is necessary to design the overall layout of the indoor LED according to the interior space and the facilities, so as to avoid the formation of the blind zone, so as to make the indoor light intensity distribution as uniform as possible. Generally speaking, the more the light source array is installed, the more it can improve the sending and receiving power, and the more effectively the "shadow" effect can be improved. But different light paths make ISI more serious, affecting transmission performance. Therefore, the rational deployment of visible light source is particularly important to achieve the best balance between enhancing the intensity of light to improve transmission performance and reducing the multidiameter ISI.

9.3.2   Visible Network Switching Technology

In indoor visible optical communication system, in addition to noise interference, there are many factors such as signal refraction, natural light, and occlusion. When the mobile terminal is transferred from one light source to another, or from indoor visible light communication to outdoor wireless communication, network switching is different from the switching process under the cellular network. To ensure that the terminal can be switched horizontally or vertically in time, it is necessary to make a reasonable prediction trigger mechanism to carry out the measurement and the upper pre-switch in advance, so as to reduce the switching delay. There is certain difference between visible light communication network switching judgment index and traditional indoor wireless access network. Its location management, transformation scale of heterogeneous network switching index, and its control strategy switching need to be redesigned. Moreover, the new visible optical terminal not only brings the diversity of business, but also breaks the application scenario of a single person-to-person communication, which puts forward new requirements for switching. Therefore, we need to set the appropriate switching scales according to the perceived attributes of terminal, network capacity, and business needs. We need to achieve vertical switching between heterogeneous networks quickly and efficiently based on the forecast trigger.

In the visible light network environment, it is necessary to find new access to the base station and select the appropriate access mode when the signal of the current visible optical base station needs to be switched immediately. To avoid unnecessary repeated switching, the first step is to set the appropriate intensity of light signal for visible network switching. Before switching, the base station will track and record the optical signal of the mobile terminal and combine multiple switching factors (terminal moving trajectory and light signal distribution intensity) to decide whether to switch.

If the decision needs to be switched, this will be done by the terminal and base station in the control plane of the top level to complete the signaling interaction, to negotiate and inform the terminal to select the virtual neighborhood to entry. At the same time, the base station and the virtual community covered by the base station are synchronized in the user plane, and the wireless business is realized by the virtual community and users at the data level. The community lingering and community re-election are conducted in base station level, for mobile terminals which are active in the community, visible light micro base station of the virtual community is transparent, there is no need to monitor any wireless paging process in virtual community, thus reducing the workload of community planning, configuration, and optimization. In addition, because the virtual community does not need to maintain the control channel of the user plane for measurement and decoding, the energy consumption of UE is reduced, and the battery life of the terminal is prolonged. Mobile terminals only need to interact with the base station when they have registered, switched, or initiated the relevant process of the Acer station to make the discovery of the virtual community.

9.3.3 Optical Network Access Control

The visible light network implements multicell stratified coverage, and the terminal can be accessed in multiple ways simultaneously, which can effectively increase the system capacity, but also brings the problem of network access control.

Since the visible light communication is a channel space with a direction and up and down channels are not independent, easy to overlap in the space, we need to consider how to achieve directional uplink access in the presence of descending interference. We can study directional access protocol based on angle perception, by deducing the divergence angle and emission angle of optical communication, analyzing each communication mode (direct communication/indirect communication) coverage, and by maintaining elevation neighbor list, real-time coordinate elevation, and the communication range, to overcome the downward interference and achieve the strong directional upward access.

Different physical technologies in the visible heterogeneous network experience different channel declines. The coordinated transmission of one or more heterogeneous access links can be configured on the sending end, while a combination of multiple links on the receiving end can obtain spatial diversity and heterogeneous diversity gain, which allows different features to be channeled through different networks. The same business flow is migrated according to the entire network state, thus improving the resource utilization of the whole system. In this case, the multimode terminal requests one or more services simultaneously to multiple AP, and the indoor gateway shunts the business data at the top to the appropriate AP, which is ultimately sent to the requesting user. When the number of users increases and mobility enhances, multiple connection control is more important: We need to comprehensively consider the AP, business types, QoS requirements, and network load conditions that the user can choose to access at the moment, so as to establish (or undo) the appropriate heterogeneous access link for the terminal.

Therefore, in order to realize multiconnection control of visible and heterogeneous network, multi-AP coordination is required to realize the collaborative communication of the link layer, and it needs to be based on optical signal sensing information. For user requirements for idle state and different business shunt characteristics, a reasonable vector decomposition is carried out, which is to rationally allocate the appropriate business volume to the proper link to maximize the gain of heterogeneous network diversity.

9.4 Visible Optical Communication Integrated Chip

Visible light communication as a new broadband wireless access method with its unique advantages has great development potential. With the further maturity of visible optical communication technology, the chip used for visible light communication will inject new vitality into the whole communication chip industry. Chip size, reduced power consumption, chip cost reduction, and the diversification of chip functions will become the general trend of integrated chip design. Visible optical communication chips include integrated LED emitting arrays, integrated PIN-receiving arrays, and integrated communication chips.

9.4.1   LED Emission Array

The integration of multiple high-bandwidth LEDs into the same chip can be more powerful than a single LED, thus increasing the distance of visible light communication. LED emission array modules include drive signal amplification circuit, AC/DC coupling circuit, and LED lamp. The launch of the drive signal directly from the signal processing chip power is small, and it is difficult to allow the communication distance to reach practical application distance, so you need to amplify the signal, in order to communicate at long distance.

The method of AC/DC coupling is used in most of the high-speed visible light communication, respectively, for AC signal (carry useful information) and DC signal (used to light up LED lights) for processing, easy to control AC signal and DC signal. The control of AC signal part can achieve the purpose of controlling the transmission distance or communication rate, while the control of DC part can control the luminous intensity of LED lights, communication distance, and speed. Finally, the signal is sent by AC/DC coupled signals driving LED lamps. The above is the emission part of a LED lamp, while the LED array chip includes the above parts. The integrated module is smaller and has better performance.

9.4.2   PIN-Receiving Array

Photoelectric detector and signal processing circuits are integrated on a chip. On the one hand, it can remove the output buffer of the preamplifier and the input buffer of the main amplifier, so as to reduce the chip area and power consumption caused by the two modules. On the other hand, preamplifier input does not require external lead but internal signal lines to connect to the input end of the main amplifier chip, which can improve the working performance of the entire front-end chip. PIN-receiving array structure can get more signal from the LED emission and undertake more and more advanced processing algorithms at the receiving end. For example, the multipath reception optical signal can be received by the sorting process, which can improve the communication speed or communication distance.

Photoelectric front-end chip is an important part of visible optical communication system. The optical receiving array is mainly composed of filter, lens, PIN tube, preamplifier, filter, and main amplifier (AMP).

The visible signal is filtered by filtered light filter to remove the background clutter. By converting the signal light through the lens, the PIN photodetector can receive a wide range of receiving range and convert the visible signal to the current signal under reverse bias. The preamplifier amplifies the weak current signal and generates the voltage signal, while the preamplifier needs to be able to guarantee low noise and high gain bandwidth. The voltage signal is used to filter out the high-frequency noise by a filter and is further amplified by the main amplifier. And the output voltage signal is processed after decoding.

9.4.3 Dedicated Visible Light Integrated Communication Chip

With the rapid development of high-speed module converter and DSP technology of digital signal processor in recent years, the digital radio frequency (RF) has become the development trend in the future. The digital processing of RF signals brings flexible and convenient means for signal processing. In a digital receiving system, the RF signal received by the central station generally needs to be digitized by analog signals using a high-resolution analog–digital converter for subsequent digital signal processing.

The transmission rate of optical communication in the future is comparable to those obtained with the optical fiber communication. The high-speed data business will provide users with a more fluid and complete user experience, and this also means higher requirements for encoding and signal processing chips. Coding and signal processing chips feature high integration and high performance. The traditional special integrated circuit (ASIC) cannot be the preferred communication unit because of its inflexible hardware structure, power consumption, and cost limitation. The special integrated communication chip module of visible light is shown in Fig. 9.5. It can be seen from the figure that the functions of coding and signal

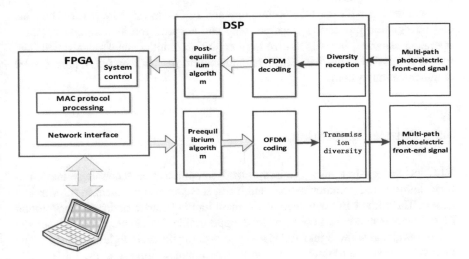

Fig. 9.5 Special integrated communication chip module for visible light

processing chip mainly include the OFDM decoding of DSP end signal, equalization processing and the MAC protocol, system and network interface control of FPGA. The development direction of the two represents the development direction of signal processing chip.

Now, DSP technology has been widely used in communications. Due to the flexibility of programmable DSP and the enhanced computing power, it will also be used in many undeveloped areas. The application of communication mainly includes software radio, speech compression coding, and GPS system. With the rapid development of VLSI, the DSP chip is developing in a faster and more stable direction while the price is falling. DSP chip changes with the application requirements. Multiple DSP parallel processing and storage structure changes to meet the chip frequency rising demand, research and development in a specific application of SOC devices will become the main trend of the development of DSP chip.

Today's digital IC is in great demand. For chip design, the usability of FPGA is not only simpler, faster, but also can save the cost of repeated flow sheet verification. Now, the FPGA technology is in a period of rapid development. The new type of chips is getting bigger and cheaper. The low-end FPGA has gradually replaced the traditional digital component, and the high-end FPGA is constantly competing for market share of ASIC. More and more processor cores are embedded in high-end FPGA chips. With the continuous improvement of semiconductor manufacturing process, the integration of FPGA will be improved and the manufacturing cost will be reduced continuously. FPGA chips with large capacity, low voltage, low power consumption, system-level high density, dynamic reconfigurable and able to integrate with ASIC FPGA chip will become the main development trend of programmable devices in the future. It is also an urgent problem to realize the miniaturization, integration, and high-speed transmission in the optical communication system.

Combination with the DSP technology and FPGA as the core part of the code and information processing chip, not only can adapt to the increasingly complex operation standard, but also realize high definition, multichannel signals, and high rate transmission. This has excellent application prospect in the visible optical communication system.

9.5   Future Expectations

VLC technology is a new wireless optical communication technology that combines lighting and communication, and it can serve as a supplement to the wireless access. The white LED will become the main lighting source under the background of the next generation, now is a good opportunitie for development. Under the background that white light LED is about to become the main lighting source of the next generation, we are now facing good development opportunities.

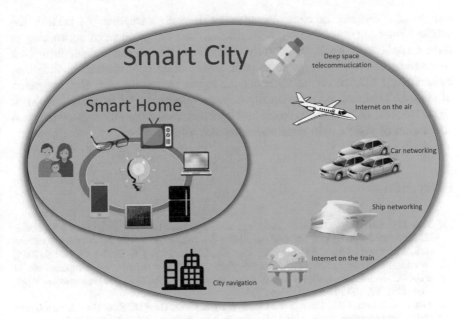

Fig. 9.6 Future prospect of optical communication

As long as there is light, the Internet access can be realized. Broadband access is the biggest application scenario in the future. Low-speed short-distance communication and positioning are the starting point of the industrialization application of visible light. Lighting network and communication network can be used to control home network and other electrical appliances connected to the grid, or even mobile terminals. Visible light communication is applied to intelligent household. Since visible technology has no electromagnetic interference compared to conventional wireless communication, it can be installed on the aircraft, including plane or satellite equipment to transfer the received satellite signal to LED lights of passenger seats, achieving light access. For long-distance optical communications, remote communication transmission information is less, which can be achieved in outdoor positioning system, such as intelligent transportation system, real-time broadcast, and identification of traffic information (Fig. 9.6).

9.6 Summary

This chapter focuses on the future development trend of visible light communication technology, mainly about the speed of visible light communication and the distance of transmission. Surface plasma LED can improve LED modulation bandwidth and increase LED luminous efficiency. The use of MIMO structure for visual imaging communication can double the speed of communication and improve the distance of communication by diversity reception. At present, the visible light communication has been realized, and the problem of networking in further application needs to be fully considered. At the same time, there is a shortage of dedicated integrated chips in visible light communication, including transmitting LEDs, receiving PINs, and the signal processing part. The integration of chip, LED signal transmission drive, signal processing of front-end chip, and signal recovery of back-end chip are problems that need to be solved in the practical application of visible light communication technology.

References

1. Yun, J.H., Cho, H.S., Bae, K.B., Sudhakar, S., Kang, Y.S., Lee, J.S., Polyakov, A.Y., Lee, I. H.: Enhanced optical properties of nanopillar light-emitting diodes by coupling localized surface plasmon of Ag = SiO_2 nanoparticles. Appl. Phys. Exp. **8**, 092002-1–092002-5 (2015)
2. Okamoto, K., Niki, I., Scherer, A.: Surface plasmon enhanced radiative recombination rate of InGaN/GaN quantum wells probed by time-resolved photoluminescence spectroscopy. Appl. Phys. Lett. **87**, 071102-1–071102-4 (2005)
3. Yeh, P.S., Chang, C.C., Chen, Y.T., Lin, D.W., Wu, C. C., He, J.H., Kuo, H.C.: Blue resonant cavity light-emitting diode with half milliwatt output power. In: SPIE OPTO. International Society for Optics and Photonics, 97680P-6 (2016)
4. Green, R.P., et al.: Modulation bandwidth studies of recombination processes in blue and green InGaN quantum well micro-light-emitting diodes. Appl. Phys. Lett. **102**, 091103-1–091103-5 (2013)
5. Huang, J.K., et al.: Enhanced light extraction efficiency of GaN-based hybrid nanorods light-emitting diodes. IEEE J. Sel. Topics Quantum Electron. **21**, 354–360 (2015)
6. Okamoto, K., et al.: High efficiency InGaN/GaN light emitters based on nanophotonics and plasmonics. IEEE J. Sel. Topics Quantum Electron. **15**, 1199–1209 (2009)
7. Sun, G., et al.: Plasmon enhancement of luminescence by metal nanoparticles. IEEE J. Sel. Topics Quantum Electron. **17**, 110–118 (2011)
8. Lin, C.H., et al.: Modulation behaviors of surface plasmon coupled light-emitting diode. Opt. Exp. **23**, 8150–8161 (2015)
9. Huang, K., et al.: Top and bottom emission enhanced electroluminescence of deep-UV light-emitting diodes induced by localised surface plasmons. Sci. Rep. **4**, 4380-1–4380-7 (2014)
10. Cho, C.Y., et al.: Surface plasmon enhanced light emission from AlGaN based ultraviolet light-emitting diodes grown on Si (111). Appl. Phys. Lett. **102**, 211110-1–211110-5 (2013)
11. Lee, K.J., et al.: Enhanced optical output power by the silver localized surface plasmon coupling through side facets of micro-hole patterned InGaN/GaN light-emitting diodes. Opt. Exp. **22**, A1051–A1058 (2014)
12. Zhu, S.C., et al.: Enhancement of the modulation bandwidth for GaN based light-emitting diode by surface plasmons. Opt. Exp. **23**, 13752–13760 (2015)
13. Lin, Y.Z., et al.: Comprehensive numeric study of gallium nitride light-emitting diodes adopting surface-plasmon-mediated light emission technique. IEEE J. Sel. Topics Quantum Electron. **17**, 942–951 (2011)
14. Langhammer, C., et al.: Localized surface plasmon resonances in aluminum nanodisks. Nano Lett. **8**, 14611471 (2008)

15. Tateishi, K., et al.: Highly enhanced green emission from InGaN quantum wells due to surface plasmon resonance on aluminum films. Appl. Phys. Lett. **106**, 121112-1–121112-6 (2015)
16. Komine, T., Nakagawa, M.: Fundamental analysis for visible-light communication system using LED lights. IEEE Trans. Consum. Electron. **50**(1), 100–107 (2004)
17. Kahn, J.M., You, R., Djahani, P., et al.: Imaging diversity receivers for high-speed infrared wireless communication. Commun. Mag. IEEE **36**(12), 88–94 (1998)
18. Wang, Y., Chi, N., Wang, Y., Tao, L., Shi, J.: Network architecture of a high-speed visible light communication local area network. IEEE Photonics Technol. Lett. **27**(2), 197–200 (2015)

Printed in the United States
By Bookmasters